The Ocean World of Jacques Cousteau

Part I:
The Art of Motion

The Ocean World of Jacques Cousteau

Part I:
The Art of Motion

Flipping their tails from side to side and using their fins for stability, these soldierfish cavort among coral growths. By one means or another, at one speed or another, most creatures of the sea enjoy a three-dimensional freedom of movement.

The Danbury Press
A Division of Grolier Enterprises Inc.

Publisher: Robert B. Clarke

Production Supervision: William Frampton

Published by The World Publishing Company

Published simultaneously in Canada
by Nelson, Foster & Scott Ltd.

First printing—1973

ISBN 0-529-05073-0
Library of Congress catalog card number: 72-87710

Printed in the United States of America

Project Director: Peter V. Ritner

Managing Editor: Steven Schepp
Assistant Managing Editor: Ruth Dugan
Senior Editors: Donald Dreves
　　　　　　　　Richard Vahan
Assistant Editors: Jill Fairchild
　　　　　　　　　Sherry Knox

Creative Director and Designer: Milton Charles

Assistant to Creative Director: Gail Ash
Illustrations Editor: Howard Koslow

Production Manager: Bernard Kass

Science Consultant: Richard C. Murphy

Typography: Nu-Type Service, Inc.

Contents: Part I

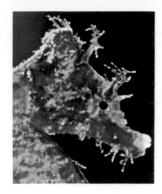

similar to the first steps of the remote ancestors of land vertebrates—and his fins may reflect the origin of our limbs.

Some creatures have chosen OTHER WAYS TO GO (Chapter VII): the waggling iguana, the surprisingly graceful walrus, the webbed-footed birds, the flying mantas, the paddling turtles—and even those coelenterates which create sails for themselves in order to be blown about the sea. Men have learned from nature, and have access from their undersea crafts to animal technology—sometimes with more power, but always with a lot less maneuverability.

An attractive way of life is that of HITCHHIKERS (Chapter VIII). As is also the case with human hitchhikers, some of them underwater are dangerous—parasites that feed on the blood and flesh of their hosts, in extreme cases killing them. Others, like the remora, just ride along doing no injury to and conferring little benefit on the larger animals.

When the flying fish launches himself into the air it is no game: he is escaping from some predator unseen beneath the surface, literally GETTING THE HULL OUT OF WATER (Chapter IX). This has the obvious advantage of reducing drastically the hydrodynamic drag. Many fish, like the mullet, leap for survival; out of the water they are temporarily invisible to whatever is chasing them. The great mammals, whales as well as sea lions and seals, leap for play, or to "spy hop," or to grab a breath of air. Man imitates the flying fish with his Hovercrafts and hydrofoils.

Modern man, of course, is always aiming TOWARD HIGHER SPEEDS (Chapter X). And most of his best clues have come from the sea: turbulence-damping shapes and surfaces, combination of modes of propulsion, propellers acting as rotating fins, paddles resembling flippers, airborne ships and jet propulsion.

The art of motion also demonstrates that MOVEMENT SHAPES A MODE OF LIFE (Chapter XI). Animals which hunt their food in the open sea tend to collect the equipment, and hydrodynamic form needed for great speed. Animals that crawl along the bottom are fitted with clumsier, but more heavily armored forms.

The need for MOVING ELSEWHERE (Chapter XII) has produced, over more than a billion years, a multitude of near-perfect designs, either fitted to or generating about all conceivable behaviors.

Introduction: Life Moves

One of the cliches of the horror filmmaker is the mobile vegetable, especially if it is carnivorous! Most of us are willing to accept—not perhaps as casually as we ought to—the contortions of serpents, the eight-legged hops of tarantulas, the multi-oared progress of centipedes, the sinister creep of the brittle star. But that a carrot should painfully pull itself up out of the ground and wriggle towards us: that is terrifying. It was, after all, the ambulatory wood of Birnam which persuaded Macbeth that all was lost. Plants should stay put, our instincts tell us. Movement is for animals!

Motion is such a critical component of animal-ness that it deserves, and gets, a huge body of special research. And this study has yielded some particularly enlightening results in the undersea world, because in the ocean the general challenges of propulsion are complicated by the density of the medium through which the motion must be made. In the sea we find not one but three "tops of the line," three elaborate near-perfect animal machines—each employing a different mode of propulsion. First, the ultimate jet-propelled creature: the giant squid. Second, the ultimate streamlined cruiser: the tuna. Third, the ultimate high-efficiency thermodynamic machine: the dolphin.

Where did these three lines of development originate? How did propulsion come to be a feature of the Animal World, how does it fit into the evolutionary picture? It is not possible to answer these questions with certainty. But we may find clues by imagining the life-style of an extremely primitive one-celled creature, looking much like our familiar contemporary, the amoeba. This infinitesimal proto-animal must eat. As he wraps his protoplasmic arms around an even smaller prey, or after digestion rejects the unusable excrement, tiny spasms ripple over his cellular membrane imparting an impulse to the little body. He begins to drift through the watery wastes. Soon he finds himself in another part of the sea with fresh food supplies. Again he feeds, again his cell walls convulse, again he forges on to new regions.

In other words, the act of nutrition itself may imply propulsion in animals. As the millennia and the millions of years pass by, those amoeba-like animals which move sluggishly, and therefore encounter less food, tend to die out. Those that move faster find more extensive grazing and tend to survive. Later another theme joins the evolutionary tale. Those animals which learn to *sense* a food particle, and to move *purposively* toward it, not only survive but set foot on the path leading to nervous systems and the manifold differentiated forms of animal life we find in the fossil record and around us on earth today. Those animals that can move but that develop little if any motivated control of their movements—which means that they must count on chance to carry food to them—tend to die out or to invest in alternate strategies of species survival: i.e., sheer fecundity.

All really efficient systems of propulsion are found in the ocean. In open water the two main mechanical ones are the jet system and the body/fin/undulation system—represented by the squid and the tuna. In addition, countless minor systems have been spun off to cope with other styles and niches of life: crawling, walking, jumping, burrowing, etc., etc. In the case of the third "top of the line" animal—the dolphin—superb propulsion equipment is supplemented by related systems we will discuss in detail later: a complex skin structure that dampens out the turbulence the dolphin's velocity creates and the mammalian warm-blood physiology that enables the dolphin to function with exceptionally high efficiency across a wide range of temperatures.

It is fascinating to trace the parallels between the naturally evolving modes of propulsion in the sea and man's artificial ones. Man swims, but with ridiculous clumsiness as compared to any fish. Yet the skindiver's flippers are surrogate fins. The wet suit is the analogue of the dolphin's and sea-lion's insulating skin and blubber layer. The jet principle, of course, man reserves mainly for travel through the air. Our engineers have not achieved the skills necessary for constructing a vehicle which moves by undulation, any more than they have built a machine which truly flies as does a bird. Instead, to propel most surface ships and submarines we have coupled a rotating fin—the propellor—to highly inefficient forms of power like the reciprocating steam-engine or the internal combustion engine. Modern nuclear plants are also driving ships with propellors. Jet units are only used on sophisticated speedboats and hovercraft. In terms of shapes our borrowings from the sea have come easier. The same hydrodynamical principles that over the ages have polished the streamlined forms of the great fish are applied in designing hulls, rudders, keels, or planing surfaces.

One of these days man will voyage from solar system to solar system in vessels powered by gigantic sails that catch and direct the ions flowing out into space from all the active stars. In theory ionic propulsion is capable of speeds very close to the speed of light. Even this has its underwater inspiration. One of Albert Einstein's earliest scientific papers concerned Brownian motion—that random, never-ceasing dance of microscopic particles in a fluid as they are struck by the molecules of the fluid. Here again, at the outset of the career which transformed our world, the sea provided the clue.

Jacques-Yves Cousteau

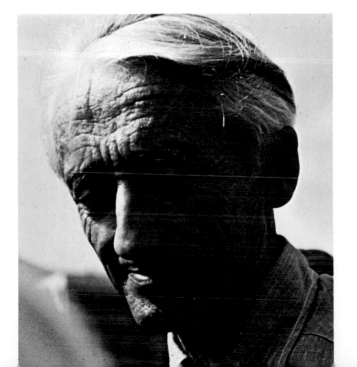

9

Chapter I. The Dense Medium

Until we engage in some strenuous physical activity, most of us never think about how much effort and energy we use in just moving around. Land-dwelling animals, like ourselves, constantly fight gravity—the force of the earth that untiringly pulls everything toward its center. When we climb a flight

> "In water a buoyant force pushes objects toward the surface, nearly balancing the force of gravity."

of stairs, we lift our entire weight with each step. When we walk across a room, our legs and feet must lift our weight and push it off the floor. Even standing still doesn't free us from gravity's bond—our legs must continue supporting our weight and thousands of small muscles around our veins must push the blood up toward the heart. While we must constantly exert ourselves to remain upright, we have no difficulty moving at low speeds through the air surrounding us. Seawater, on the contrary, is 800 times more dense than air and greatly hinders movement through it, as you know if you have ever tried to walk through waist-deep water.

The water in the oceans is the same as that in lakes and streams, but ocean water has a greater quantity of materials dissolved in it. Most significant of these are the salts. Salt increases the density of water, so ocean dwellers have a slightly more dense medium than their freshwater counterparts.

Viscosity is the internal frictional resistance of a substance, or a measure of the attraction of its molecules to one another which creates a tug on bodies attempting to move through the substance. Try drawing a spoon through a jar of honey and then through a container of water. The spoon passes through the water more easily than through the honey, because the honey is more viscous than water. But to make a fair comparison of relative viscosities and their influence on the movement of animals we should compare the viscosities of water and air, the two mediums that can sustain life.

Ocean water has a greater quantity of salts and materials dissolved in it than in lakes and streams. Salt increases the density of water, so ocean dwellers live in a slightly more dense medium than their freshwater counterparts.

In spite of the problems involved in moving through a dense and viscous medium, marine animals do enjoy a certain advantage over land-dwellers. In water a buoyant force pushes objects toward the surface, nearly balancing the force of gravity. Because of this buoyancy, in water the specific weight (density) of a submerged object is reduced: a piece of wood will float; a squid will move around practically weightless; a three pound aluminum ingot will weigh only two pounds; a seven pound pigiron, only six pounds.

Supporting world of whales. The majestic California gray whale parades up and down the west coast of North America on an annual 5000 mile march. Smaller by about half than its enormous cousin, the blue whale which is the largest animal ever to inhabit the earth, the gray still weighs in at an impressive 40 tons and measures up to 50 feet in length. Several million years ago the ancestors of the whales adopted water as their living space and grew to mammoth size. In water they are freed of the force of gravity; the buoyancy of water supports their massive bodies. Out of water whales must die. When they are stranded on a beach, their lungs collapse under the great weight of their bodies and they suffocate.

Logs *float because the volume of water they displace weighs more than they do.*

Buoyancy

In the third century B.C. Archimedes was the leading scientist at court of Hiero II of Syracuse, a Greek city in Sicily. Legend holds that the king was in the market for a new crown, one made of pure gold. A metalworker offered one to the king for an extraordinarily low price. The king suspected that this crown was not made entirely of gold and thought that silver may have been substituted. But, having no way to confirm his suspicions, he called in Archimedes to prove his theory.

Archimedes took samples of pure silver and gold, which were equal to the crown's weight. He immersed each in a water-filled container and measured the amount of water that overflowed. He discovered that the volume of water displaced by the gold sample was less than that displaced by the silver. When the crown was immersed, the volume of water that spilled from the container was less than the amount displaced by the silver, but more

> "Archimedes's principle: An object immersed in a liquid is buoyed up by a force equal to the weight of the volume of liquid it displaces."

than that displaced by the gold. Archimedes concluded that the crown was a combination of gold and silver, and the attempted deception was uncovered.

Archimedes' experiment demonstrated for the first time the buoyant influence of water, and the principle underlying this phenomenon was named after him. Archimedes' principle states that an object immersed in a liquid is buoyed up by a force equal to the weight of the volume of liquid it displaces. An object floats when the buoyant force exceeds its weight. It sinks when the displaced volume of liquid weighs less than it does. If an object displaces a volume of liquid that weighs exactly the same as it does, it is said

Heavier than water. After centuries on the bottom of the sea, this heavy iron cannon is encrusted with marine life, and corroding away. In the sea it weighs less than it does on land because of the buoyant force of the water. The difference between the cannon's weight on land and in the sea is the weight of the water it displaces, in this case one seventh of its weight in air.

to be neutrally buoyant; it neither sinks nor floats. Such neutrally buoyant objects are currently built in plastic to drift at specific depths and are used to measure deep sea currents.

Two gulls soar in over a beach to join others already standing near the water's edge. When coming in for landings, gulls twist their wings in very much the same way an airplane lowers its wing flaps.

To Defy Gravity

The effortlessly soaring gull has not escaped gravity but has temporarily lifted itself into a supporting airflow. It had to beat its wings with great force and rapidity to reach the invisible cushions of air on which it rides. If the updrafts fail, the bird must again flap its wings to stay aloft, or it will sink to earth. Although its anatomy is almost perfectly suited for aerial life, a bird must expend tremendous amounts of energy to become airborne. A tiny bird burns at least one percent of its body weight for each hour of flight, and an active small bird must eat almost its body's weight of food each day. It is costly to

These cardinalfish are found hovering above the bottom on coral reefs. With the water to buoy up their tiny bodies they can remain there effortlessly and consequently need very little food.

fly in the face of earth's gravity. To reduce that cost the central temperature of a bird is high in order to increase the energy output of the muscles, its skeleton is thin and hollow, and streamlining is achieved with feathers.

Sea birds, however, have to find their food in the sea; the stormy petrel hovers close to the surface and picks its small prey from the first inches of water; boobies and terns dive bomb for short, fairly shallow but efficient intrusions; cormorants can stay five minutes in the sea and "swim" half a mile with their wings. Heavier than air, birds are lighter than water, and must fight their way down into the sea, as they fight their way up.

Swim Bladder

The ocean is a three-dimensional world for all marine creatures. Some of these are sedentary, while others constantly travel in the open sea. Rock and reef fishes need not cover great distances; most of their life is spent moving slowly about in the vicinity of their home. They are greatly helped in their command of the third dimension by a built-in

> "In a rapid ascent, a fish must eliminate gas from its swim bladder very quickly to avoid ballooning."

gas-filled chamber, or swim bladder: without it, they would slowly but constantly sink to the bottom, because their bodies, made of flesh and bones, are slightly heavier than sea water. The swim bladder increases their vol-

Grunts with their swim bladders, like this one, can regulate their volume at will to stay hovering without effort at whatever level in the sea they want.

The angelshark seen at right is launching itself from the bottom with powerful sweeps of its pectorals. Because it is heavier than water and has no swim bladder, this ray must make constant efforts to stay above the sea floor.

ume without increasing their weight and keeps them perfectly balanced.

When they descend, the gas in the bladder is compressed and more gas must be quickly generated therein in order to keep them neutrally buoyant. If they rise, some of the expanding gas has to be eliminated very rapidly to avoid ballooning to the surface with their bladders protruding from their mouths.

The mechanism of instant control of the swim bladder is not yet well understood.

No Swim Bladder

Most of the fast open ocean swimmers, fish, squids, sharks or manta rays, have no such buoyancy system as a swim bladder. They all slowly sink when they stop swimming. This is rather convenient for creatures that spend a substantial part of their lives resting or hiding on the bottom: rays, skates, electric rays, and nurse sharks often remain motionless for hours on the sea floor. Their apparent weight, being roughly one-twenti-eth of what it would be in air, is enough to anchor them against the drag of currents, tides or swells.

For tuna, squids and most of the sharks, however, a constantly negative buoyancy is a dominant factor that imprisons them in a way of life: crossing oceans for weeks or months at a time at the depths where their food can be found, but with miles of water under them, they are sentenced to swim forever and probably never to sleep.

Chapter II. Form and Design

From a general viewpoint, in order to move, an animal must generate lift to counteract gravity, and must generate thrust to counteract drag. We have seen that fish are privileged because they are buoyed up by the water's lift and thus have to make no effort to overcome gravity. On the other hand, they are handicapped because they need to develop very substantial thrust forces to penetrate swiftly the oceans.

An Apollo spacecraft hurtles through the vacuum of space, free from the forces of gravity and from drag. All the launching's rocket energy was spent to snatch the craft from the pull of the earth and from the resistance of the atmosphere. Now, in outer space, it can maintain its speed at no expense of fuel. Its appendages stick out here and there; its shape does not matter because out of the atmosphere there is no drag, no turbulence to slow down the vehicle. But in air, propulsion obeys the laws of aerodynamics; modern aircraft design has been inspired by the shapes of such birds as the swallow or the albatross. The streamlining of a supersonic jet plane is sleek, its shape being studied in every detail to reduce turbulence and offer the least possible resistance.

In water, the laws of hydrodynamics are generally similar but much more severe, because of the high density and high viscosity of liquids. Aquatic organisms experience a very strong drag, which is a function of speed, of shape, of the surface of the body, and of the nature of the water flow over it. These four factors cannot be considered separately because they interefere heavily with one another. The drag forces become very large at rapid speeds because they increase approximately as the square of the velocity. A fusiform shape, with the thickest portion one third of the way back, like a tunafish or a torpedo, tapered at either end, separates water easily in front and allows it to converge smoothly behind. In water as in air, it is the best streamlined design; the higher the speed, the more elongate the shape must be with a reduced cross-section, but also with an increased surface to encompass the same volume of flesh. Now, when the surface of the body increases, the drag due to friction increases, and the remedy is to improve the nature of the surface, as skiers do when they wax their skis. Finally, any structures projecting from the body tend to produce turbulence, so even fins and flippers are streamlined and eventually can be folded against the body or even inside special slots on the back. Even the inside of the mouth and especially the gills are well streamlined to reduce the drag of "inner water flow."

The basic fusiform shape may be compressed either in the vertical plane or in the horizontal plane. Those marine animals that live a sedentary life don't need speed; their shapes are not governed by hydrodynamics.

Underwater disturbances — slow movement. The wake left behind by this swan has created turbulence in otherwise still water. As the swan moves it forces water ahead, setting up wavelets on the surface and disturbances beneath the surface. These underwater disturbances, turbulence, whether made by birds or mammals on the surface or by fishes and other animals underwater, tend to slow the swimmers, clinging to their sides, hindering their progress. In front, you can see how the water tends to furrow upward as snow does in front of a snowplow, slowing forward progress. What you can't see are the eddies, swirls, vortexes of water all around the swan. These eddies, swirls, and plowed-up water have the same effect on fishes. But fishes, living as they have in water for 350 million years, have adapted to their environment. Those that need to swim swiftly have made the necessary body and shape adaptations to cleave the water with a minimum of hindering turbulence.

The Price of Speed

When the Atlantic dolphin, shown here, breaks the surface, he creates a wake that acts as a strong drag. When totally submerged, the dolphin does better, being able to accommodate his shape to practically eliminate turbulence around his body. At slow cruising speeds, the surface effect does not influence noticeably the economy of the dolphin's propulsion. But at high speeds, the dolphin jumps clear out of the water to breathe, to reduce the time during which he pulls a costly wake behind him.

Two types of drag can reduce the speed of marine animals, large or small, in the water: "pressure drag" and "friction drag." Pressure drag is a direct function of streamlining and design. In friction drag, which mainly depends on the smoothness of the skin, some water is moving with the fish against water further away which is not moving, thus creating eddies and turbulence. Fast swimmers are covered with a slimy mucus secretion.

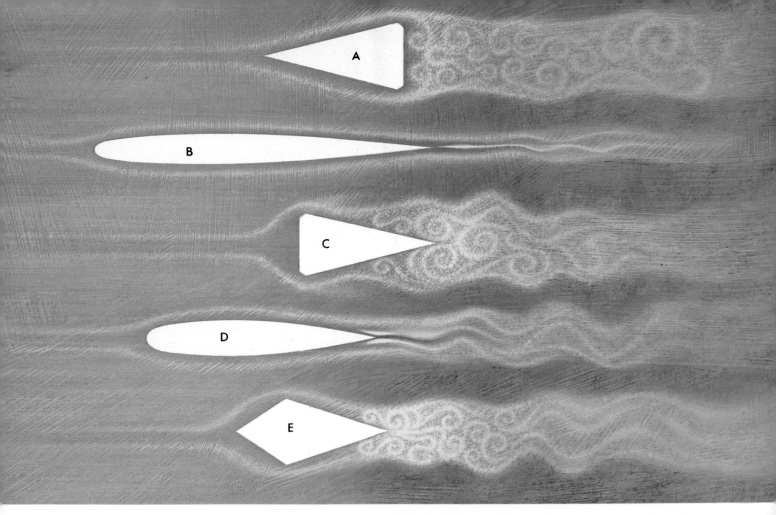

Streamlining

Water flow can be *laminar* (smooth), *turbulent* (irregular), or *transitional*. Laminar flow creates the least drag, and turbulent flow the most. In practice, any fish or mammal or man-made hull produces transitional flow, as perfect laminar or totally turbulent flows are never encountered.

Streamlined bodies promote laminar flow and the faster an object travels through water, the more streamlined it must be.

Turbulence begins to set in when the thin layer of water immediately next to the moving body (the "boundary layer") becomes unstable. Such instability is, at least partially, inevitable in man-made rigid hulls. If the turbulence in the boundary layer can be stabilized, then laminar flow can be maintained over the entire organism. It seems

A / Pointed nose. This shape eases cutting through the water, but the broadness of the after portion causes heavy drag and boiling up of water just behind the body.

B / Long and slim. This characteristic bullet shape of the fusiform body like that of a barracuda or a shark. It creates less turbulence as this drawing based on a photographic study indicates.

C / Blunt front. This is the least efficient shape shown; it stirs water and creates the most turbulence.

D / Shortened and broadened. This shape creates greater turbulence than does the fusiform body.

E / Angular on the sides. This shape creates still greater turbulence.

that fish and marine mammals have such a stabilization mechanism: by constantly changing their shape to conform their body surfaces to the lines of flow they are able to move at speeds that could not be matched by exact, but rigid, replicas of their forms.

21

Fish Shapes

Besides displaying a wide variety of sizes and colors, fish come in many shapes. Where the fish lives, how it feeds, what it eats, how fast it swims, and what its relationships to other animals are — all may be affected by its body shape. The shape of fish's bodies fall into general categories: fusiform, like the shark, barracuda, and codfish; laterally (side to side) compressed, like the angelfish, spadefish, and filefish; dorsoventrally (top to bottom) compressed, like the skate and guitarfish; attenuated, like the conger or the American eel; and a number of other shapes that might be called miscellaneous because there are only a few examples of each (these might include the strangely shaped seahorse, the triangular cowfish, and the globular porcupinefish). Whatever their shape, fish are similar in their bilateral symmetry.

Although we have given general categories of fish shapes, most species exhibit characteristics of more than one category. Few fish, for example, are precisely tubular, thus fitting the exact definition of a fusiform fish. Most that come close to this shape are usually somewhat flattened dorsoventrally or laterally. A few are long and drawn out as well as tubular and therefore combine fusiform and attenuated shapes.

The fish has assumed its present shape through many millions of years of natural selection. That is, the individuals of each species best suited for their particular environment had a better chance to survive long enough to reproduce and pass on their genetic material to their offspring, who then did the same. Those less suited either moved to more suitable environments or died before reproducing and passing their genes to offspring.

Laterally flattened. The blue- and yellow-striped angelfish seen here are more laterally flattened than most other fish. It allows them to retain their swimming speed and gives them the additional advantage of fitting into narrow crevices, whether seeking protection or food.

Their near relatives, the butterflyfish are flattened this way too. The spadefish, a white and black striped schooling fish of the tropical seas, takes the same form.

Fusiform and laterally flattened. Soldierfish such as these are one of the many species that combine fusiform and laterally flattened shapes. They dwell in the open areas of reefs and come out frequently at night.

The squirrelfish, which are in the same family, share these characteristics. Another family of fishes that outwardly resemble them, the big-eyes, do too. Some of the small species of sea bass that never grow to more than four or five inches in length tend to be of this shape as do the tiny cardinalfish and several species of wrasses.

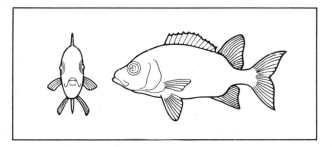

Fusiform. This blue-spotted rock cod, a sea bass found on Australia's Great Barrier Reef, comes close to the torpedo shape of fusiform fish. When it must, the sea bass can move at express train speed to strike its prey.

Other members of the sea bass family share the shape of this fish. So too do the billfishes, including the several species of marlin, the swordfish and the sailfish, which exceed the sea bass in speed. Lizardfish strike their prey with reptilian swiftness.

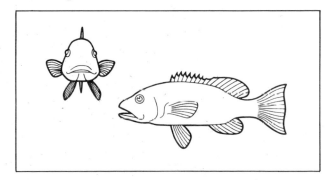

23

Attenuated and cylindrical. Trumpetfish, like this one, are attenuated in shape and cylindrical in general form. They are not as flexible as the eels or morays, but their proportions are rather eellike and enable them to mimic some of the staghorn corals they normally dwell among.

The rarer cornetfish resembles the trumpetfish strikingly, differing chiefly in having a long filament extending from the tailfin. Pipefish, who dwell in the eel grass and manatee grass of tropical and temperate coastal areas, match these plants with their long, drawn out and cylindrical shapes.

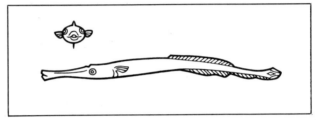

Attenuated. The moray pictured here, a variety of eel, has an attenuated shape, which enables it to slip in and out of openings in the rocky reefs it inhabits. Having a small diameter relative to its length, the moray is able to seek food in the small holes that honeycomb many reefs. It frequently lies in a dark hole waiting for a potential prey to pass by. Unwary divers have thrust hands into holes and have been bitten, but morays are usually not aggressive.

The morays are only one of more than two dozen families of eellike fishes in the same classification that have an attenuated body.

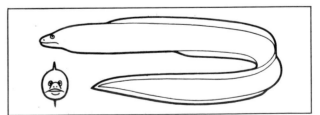

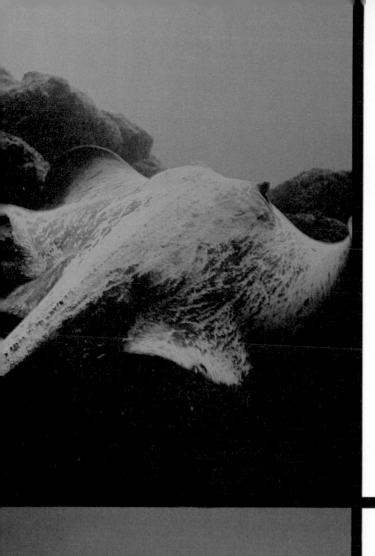

Dorsoventrally compressed. The flattened form of this stingray enables it to move along the bottom almost surreptiously. Flattened top and bottom, it is the epitome of the dorsoventrally compressed form. In adapting to life on the ocean floor, rays, and their relatives the skates, have become flattened and have developed a unique means of locomotion dependent on their body shape.

So too have their relatives the sawfish and the guitarfish. A shallow water angler, the goosefish, is another dorso-ventrally compressed fish that dwells almost exclusively on the sea floor.

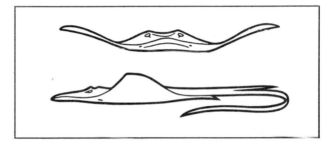
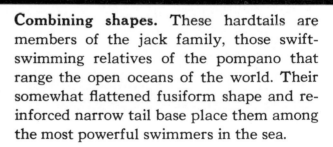

Combining shapes. These hardtails are members of the jack family, those swift-swimming relatives of the pompano that range the open oceans of the world. Their somewhat flattened fusiform shape and reinforced narrow tail base place them among the most powerful swimmers in the sea.

Other swift swimmers that fit this category include the tuna, mackerel, and mako. In addition the striped bass of the north temperate coastal waters and the bluefish of the same areas have a combination shape. Parrotfish of the tropical coral reef areas of the world do too.

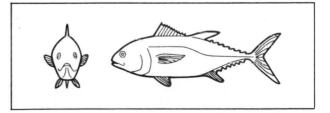

Cetacean Shapes

The cetaceans, which include the whales, dolphins, and porpoises, have adapted to a totally aquatic life since their ancestors returned to the sea nearly 70 million years ago. Before that time the toothed whales were meat eaters and the baleen whales were probably ant eaters. Living by the edge of the sea, they could find meat in the form of fish, squid, or shrimp. And as they entered the sea from shore in search of food fish, they began to adapt. Such evolutionary trends occur today when some modern fishermens dogs develop webbed feet and dive to 10 feet.

The most important changes were those having to do with the way the animals moved and breathed. They reassumed the fusiform shape of early fish. The bones in their necks became shorter until there was no longer any narrowing between head and body. With water to support their weight, they became rounded or cylindrical in body shape, reducing the drag irregularities. Front limbs adapted by becoming broad, flat, paddlelike organs. But internally, cetaceans still retain evidence of fingers. X-rays reveal up to 12 extra finger bones in some species. The hind limbs disappeared, leaving only a trace internally that there ever were any. The tails developed into flukes. It is the flukes, combined with the powerful muscles of the trunk of these animals, that provide the propulsive power enabling them to swim and dive so efficiently. Unlike the fishes, cetaceans' flukes are horizontal, moving up and down. In the fishes, tail fins or caudals are vertical and move from side to side, providing an easy-to-recognize differentiation between fishes and mammals in the sea.

Today's whales, dolphins, and porpoises are born, live their lives, and die in the sea. They

are totally aquatic. Their movements are appropriate only for the aquatic environment. Their streamlined forms and powerful flukes enable them to swim with ease. The big whales have grown to be the largest animal that ever lived, outweighing the biggest dinosaurs 3 to 1. Such large structures could not survive on land—because of their body weight they would suffocate.

Another change the cetaceans underwent in adapting to their re-entry to the sea was the position of their nostrils. From a position on

> "These returned-to-the-sea mammals are voluntary breathers, breathing only upon conscious effort—unlike man."

the upper jaw as far forward as possible, the nostrils moved upward and backward until they are today located atop the head, sometimes as a single opening, sometimes as

a double opening. And these returned-to-sea mammals became voluntary breathers, breathing only upon conscious effort unlike man and other mammals who are involuntary breathers.

The development or return of a dorsal fin for lateral stability was another change that took place in some of the cetaceans upon their return to the sea.

A / Porpoises. *Tapering profile and jaws integral with skull.*

B / Sperm whale. *Enormous square head and heavy body.*

C / Dolphin. *Tapering head and beaklike jaws.*

D / Gray whale. *Heavy bodied with mouth high on head for filter feeding.*

E / Narwhal. *Has a single upper incisor that resembles the mythical unicorn's horn.*

F / Humpback whale. *Tapered in front and back, but thick-bodied between very long flippers.*

Imitating Sea Animal Forms

Man has taken many lessons from animals that inhabit the sea. Our imitations of animal forms show in aircraft, submarines, blimps, surface craft and automobiles. We seek to imitate the streamlined form of fish like the tuna, the mackerel, the shark and other fusiform fish which are built for speed. Airplanes and submarines, for example, have cylindrical bodies, tapering at the ends. Both have broad plane surfaces analogous with a fish's pectoral fins. In the airplane it is the wings. In the submarine, it is the diving planes. Both have a vertical plane analogous to either the dorsal fin or the caudal fin of the fish. In the plane it is the vertical tail assembly. In the submarine it is the 'sail,' the part of the boat that houses the periscopes and conning tower jutting above the

The racing motorboat, above, a flat plane that skitters over the water's surface, seems to fly. Its pointed bow helps it knife through the water as it gathers speed to lift up out of the water. The shallow keel, which runs the length of the otherwise flat-bottomed hull, gives the boat stability.

*At right, the **Atlantic dolphin** with its torpedo-like body speeds along, sometimes leaping clear of the water as it goes. The speed it attains enables it to make these leaps—like the motorboat that breaks free of the water. And it is during these hurdles that the dolphin can breathe without breaking stride.*

pressure hull. Our surface craft are handicapped by the fact that they move half through water and half through air and that their design is a compromise between good streamlining and seaworthiness. But of course there is no model of a wheel in the animal world, and it is the privilege of man to have materialized the axis of a spinning body. Once the wheel was invented, it was inevitable that paddlewheels and screws

would be used to push various types of hulls through water. Modern ships' propellers are really just an assemblage of several rotating fins. Slow ships like tugs have huge propellers, each blade having hydrodynamic characteristics comparable to those of a grouper's tail, while speedboats have small, rapidly rotating screws, with short and broad blades like the caudal fins of tuna.

Comparing all animals and all man-made vehicles moving through air, on land or underwater, the most efficient will be those that require the least energy to transport one pound of matter over one mile. By far the best result is obtained by man on a bicycle, second best are open ocean fish and sea mammals, then horses, then jet airplanes ... way behind come helicopters; bees and mice trail the list. Ironically, the simple bicycle is the only invention of man that beats marine animals in efficiency.

Chapter III. Traveling Wave

When a fish hanging motionless near the bottom is disturbed, it bursts from its resting place with only a few rapid strokes of its tail, fleeing possible trouble. It would seem that a fish should have great difficulty accelerating rapidly in a medium as dense as water. Some of the energy is spent to generate acceleration; the rest is spent to counter friction and

> "Fish reach their maximum speed by flicking their tails a few times. How do they achieve this great thrust?"

pressure drag, which are roughly proportional to the square of speed. Yet fish seem able to reach their maximum speed by flicking their tails a few times. To accomplish this feat, they must be extraordinarily able to get good "traction" on the water around them. How can they achieve the great thrust necessary for quick movement?

The most common form of locomotion among the aquatic animals is undulation. The body is thrown into a series of curves that begins at the head and passes along the length of the body as a *traveling wave*. Among most fish these body waves move in the horizontal plane, but in flatfish and in many marine mammals the body waves move in the vertical plane. On of the most typical examples of pure traveling wave propulsion is an aquatic reptile, the sea snake, quite common in some equatorial seas.

The earliest vertebrates probably swam in a similar manner, with rhythmic muscle contractions flexing their bodies from side to side. Pushing against the water this way resulted in forward movement. Supporting

this theory is the marine iguana from the Galapagos Islands.

Nearly all fish employ one of three general swimming techniques stemming from these traveling wave movements. The first is that employed by the sea snake, and attenuated fish, like eels and ribbonfish. Their motions are serpentine; the traveling wave bends their bodies into curves that increase in amplitude and decrease in wavelength as they progress backward. Because they have only small ineffective swimming fins, if any, they entirely rely on undulations of the body to gain a "foothold" on the water.

Fish with rigid bodies, on the other hand, like the armor-plated cowfish, are unable to flex their bodies to help them swim. They use only their short tails, swishing from side to side in a short arc, to push them along.

Between these two extremes is a third method, used by the majority of fish, combining characteristics of both. The wigwagging of the caudal fin is coordinated with subtle body undulations. This method yields smooth, rhythmic motion and improved efficiency. The fish's head swings in a small arc, and the body and caudal fin curve to form a complete transverse wave. So our startled fish, if he belongs to this category, gets a powerful takeoff leaning on the water with its body as well as with its tail. Men with their rubber foot fins do their best to imitate this type of swimming.

Epitomizing the traveling wave. Startled into action, the sea snake gives a perfect example of the traveling wave. Distinct S-curves travel from the animal's head and along its flattened body, and in increasing breadth, finally reach the tail. Studies of the eddies created at each curve of the body indicate that the supple snakes may actually roll on the swirls they cause in the water.

The Living Box Moves

The inflexible armored body of the trunkfish is fine protection against some of the predators it must occasionally face. But it renders the strangely shaped fish awkward and slow moving. In cross section, the trunkfish and its near relative the cowfish are either triangular or rectangular, and their shapes further contribute to their awkwardness. These animals in armor are poorly equipped to outswim an enemy. Under normal cir-

A spotted trunkfish sculls about the coral reef in search of food. Its rigid body makes it a clumsy swimmer, but in an emergency it can usually escape approaching danger fast enough.

cumstances, with no cause for alarm and no imminent danger, the trunkfish sculls about with its transparent, fan-shaped pectoral fins. Its equally small, delicate dorsal and anal fins contribute a wave form motion in this normal mode of swimming. Its caudal fin barely moves. Frequently, the fish hangs in the water on the reef, barely moving any of its fins. It remains near the bottom or close to cover of some sort. When danger approaches, the awkward trunkfish and cowfish flail about to reach cover. Their caudal fins lash violently from side to side on the hinge at the base of their tails—the peduncular hinge marked with a sharp spine that offers additional protection against attack from behind. But the main thrust of swimming comes from a special way of moving developed over many thousands of generations. The accompanying photograph and chart show how the trunkfish and cowfish move to meet these occasional life-and-death emergencies.

This unique method of swimming is called ostraciiform movement. It is peculiar to the trunkfish, the cowfish and two other fish families, the puffers and the porcupinefish which are related to them. The name for this type of swimming comes from the family name (they are variously called Ostraciidae and Ostraciontidae) of the trunkfish and the cowfish.

In the trunkfish and the cowfish it is the bony armor that makes ostraciiform swimming necessary. Among the puffers and porcupinefish that use the same movement, the need arises from their awkward shape.

The trunkfish's bony armor is best described as a living box of polygonal bony plates. Some of the plates have minute spines and tubercles giving their surface a rough feeling. One curious habit some divers have observed in trunkfish and cowfish is that of blowing jets of water at the fine sediments on the sea floor around the reefs. They probably use these jets in their search for the minute organisms they eat. Each time they use these water jets, the sand in front of them billows up in clouds bringing up a few tiny organisms. At the same time, the jet pushes the fish back an inch or two. The

visual effect is a back-and-forth movement which is interrupted periodically when the trunkfish (or cowfish) darts forward a few inches more than usual to suck up the small animals it has found.

A / Ready for movement. *By contracting all musculature to one side, the trunkfish puts its somewhat streamlined head and its tail fin on the same side of its median axis.*

B / Movement begins. *The fish now straightens itself and its tail and head cross the axis of progression at the same time. At this instant the tail has a strong forward thrust, and this drives the animal ahead. Pausing in this position momentarily, the fish takes advantage of the flow of water that pushes it along.*

C / Movement continues. *Once again the trunkfish contracts all its muscles, this time to the side opposite that in position* **A**, *with head and tail together on the other side of the median axis.*

D, E, *and* **F / Repeating the process.** *The fish continues the movement, alternating from one side to the other, and makes its way through the water.*

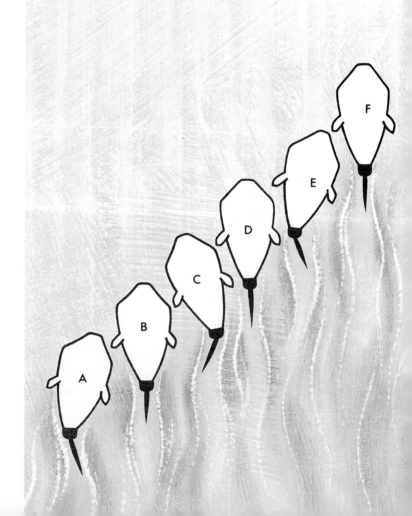

Effect of Muscle and Bone Action on Swimming

It was once thought that fish swim solely by using their tail fins, which they swept in an arc behind them. But motion-picture analysis of fish shows that fish swim by side-to-side undulations of their entire bodies. In fact, fish whose caudal fins have been amputated can still swim, some almost as well as intact fish of the same species. We know the tail fin facilitates swimming, but it is not the exclusive source of propulsion power for most fish.

The muscles along the sides of a fish are the strongest it has, and those associated with the fins are relatively weak. In swimming, a succession of contractions passes along each side of the fish. Some of the W-shaped muscle segments, or myomeres, contract on one side, and those opposite them relax and

stretch. This bends the fish's body, and the fish pushes against the water first with one side and then with the other.

The flexible frame of the fish is a good foundation for the muscles. The backbone extends from the head to the tail and is made up of many interlocking vertebrae. The vertebrae are jointed to allow side-to-side movement and are strong enough to withstand the great strain placed on them by the flexing muscles.

A / Cutaway of a salmon. *This shows the backbone, or vertebral column, of the fish and some of the numerous bands of muscle segments that gird the fish. Note that the spiny rays of the dorsal and anal fins are not attached to the backbone but simply are anchored in the flesh of the fish.*

B / The backbone bends from side to side. *Muscle contractions pass in waves along each side of the fish. While a series of muscle segments contract on one side those on the opposite side relax allowing the fish to bend at that point. The fish gets its thrust when it contracts and relaxes muscles alternately— first on one side, then the other. So the principal thrust comes not only from the caudal fin itself but also from these muscle segments or myomeres.*

35

Orca's Traveling Wave

In an easy undulating motion, the orca (killer whale) breaks the water's surface with the top of his head. A puff of vapor issues from his blowhole as he exhales. As he quickly inhales, his back with its tall dorsal fin breaks the surface too. He bows his head and points downward even as his great, broad flukes flash momentarily out of the water. And he's gone—submerged beneath the sea's surface as smoothly as if he hadn't passed that way with his multiton hulk.

Those great broad flashing flukes on orcas, and other whales and dolphins, are planes that normally are parallel to the surface of the water. They are moved up and down driven by powerful muscles in the body ahead of the tail. The muscles are connected to the tail and flukes by a series of tendons. When the orca bows its head downward to sound, its back arches and a traveling wave passes along the length of the animal's body with increasing amplitude. As it reaches that part of the body that houses the muscles that power the tail and flukes, the great ani-

Surfacing. *Bubbles of exhaled breath appear as the orca surfaces to take some fresh air.*

Fin over surface. *The dorsal fin appears above the surface with water streaming from it.*

mal "snaps the whip" and the broad flukes drive the animal forward.

It is a fortunate coincidence that the vertebral column of marine mammals developed on land by their ancestors allows them to bend in a vertical plane, so that dolphins and whales can "sound" to feed and "surface" to breathe easily and often. Fish, on the contrary, which do not need to surface, have developed laterally flexible spines.

In captivity, orcas have demonstrated some of their vast strength. They can learn to leap high up, leaving the water entirely with their great bulk. In some aquariums where they are held, a white-sided dolphin shares the pool as a companion, dwarfed in size but apparently compatible with the ineptly named killer whale. These dolphins move in virtually the same manner the orcas do. Some can "tail walk," leaping up until only their flukes are in the water, then remain in a vertical position just above the water by wagging their flukes vigorously just beneath the surface.

Sounding. *Arching its back, the orca prepares to sound. Its head has already entered the water.*

U-turn. *Dorsal high in the air, the orca is in the middle of a sweeping U-turn.*

Chapter IV. The Role of Fins

Fins are to fish what arms and legs are to men. And even a little more. Most fish have two general types of fins—median, or vertical, fins, which originate along the midline of the animal, and paired fins on their sides. The median fins include the dorsals on the back; the caudal, or tail, fins; and the anal fins on the belly just behind the vent or anus. Some fish have as many as three dorsal fins; some have two anal fins. Of the paired fins, the pectorals at each side near the head are analogous with our arms and with bird's wings, and most fish have them. The paired pelvic, or ventral, fins are located below and usually behind the pectorals.

Fins have been adapted for many purposes, but they are mainly used for propulsion,

> **"A long and soft tail gives the grouper instant thrust. The broad and short caudal fin of tuna is a high-efficiency propeller for fast, long-range travels."**

stability, steering, and braking. In some cases, as we'll see later in this volume, they have been modified for other functions. As fish evolved toward faster speeds, their fins, like feathers on an arrow, provided stability and made it possible for them to propel themselves where they wanted to go. As its head moves from side to side when a fish swims, the animal has a tendency to veer off from its forward path. To resist this condition, known as yaw, the fish erects its dorsal fin. The tendency is further reduced in fish with long, slender bodies and by the long, trailing dorsal and anal fins of some reef fish with deep, short bodies. These deep bodies keep the fish from rolling, much in the way

a sailboat's keel keeps it steady. Fish of other body structures extend their paired fins to avoid rolling. To change direction vertically, fish bring their paired fins into play. The pectoral fins act as hydroplanes to raise the nose of the fish, while ventral fins bring the rest of the body into horizontal plane.

Some open-ocean swimmers eventually have their tail fin amputated by a pursuing predator. For some time these fish are able to swim about as fast as usual, using their undulating bodies, but they soon get exhausted, and do not survive. Laboratory experiments show that for a given fish the amplitude of tail beats remains practically constant and that speed depends primarily upon the number of beats per second, which have been measured as high as 25.

The main propulsive agent for most fish is the caudal fin. A broad tail gives a rapid burst of speed from a standing start, and this is useful to a fish that must dart after a meal or away from a predator. Fast, long-distance swimmers have a very long and narrow caudal fin, which is hydrodynamically very efficient and seen in such fish as tunas, jacks, and marlins. Seahorses and some species of eels get along without caudal fins, and in skates and rays they have developed into long, spikelike projections with virtually no function in propulsion.

Maneuvering for a turn. This rockfish, a member of the sea bass family, looks like it's signaling for a turn as it extends a pectoral fin. Actually the fish is using its pectoral fin to maneuver into a turn. Giving a major assist in the turn is its tail fin, curved off to its left. The soft-rayed rear part of its dorsal, or back, fin is also curved to the left, perhaps helping in the maneuver, while the spiny-rayed portion of the dorsal remains erect, lending stability so the fish won't flop over on one side. Turning is only one of the many functions of the various paired and median fins fish possess.

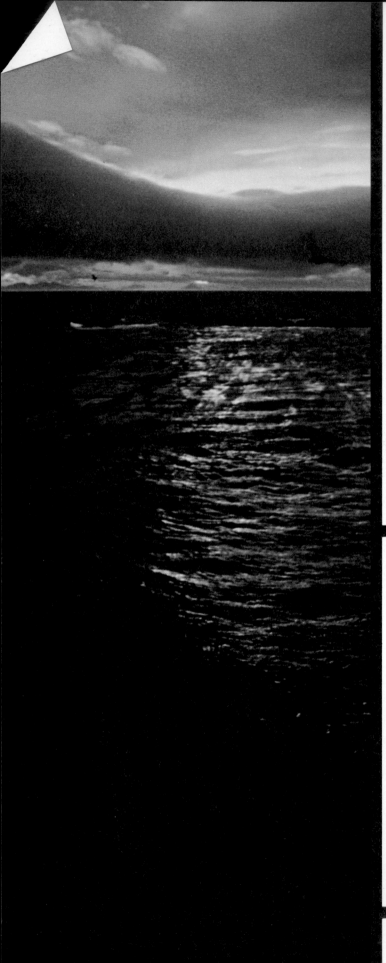

The Important Propulsive Fin

Whatever the shape of the caudal fins, their function is essential for aquatic animals. The caudal fin, or tail, is used in coordination with the massive muscle segments of the body to propel the fish through water. Most fish tails originate at the end of the vertebral column. In heterocercal (uneven) tails, the spinal column extends into the larger upper lobe of the caudal. The heterocercal tail of sharks gives these heavy fish without swim bladders an upward thrust. In homocercal (even) fins the vertebral column ends at the fin base and supports a symmetrical "tail" which produces only a forward thrust. Some homocercal tails are forked, some are square, some are rounded—but each serves a specific function. For example, the lunate (crescent-shaped) tails of mackerels, tunas, and jacks indicate fast swimmers. The broad tail of a grouper gives him the ability to accelerate very quickly.

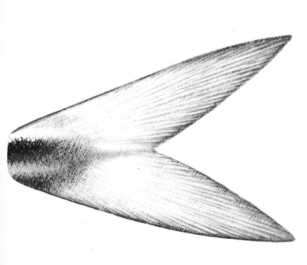

Tenpounder. *This herringlike fish with a forked tail is a fast swimmer.*

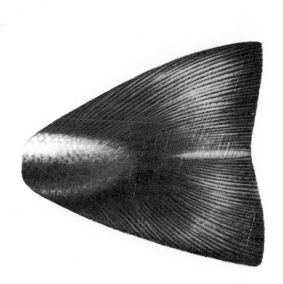

Haddock. *Close relatives of the cod, haddock are moderate swimmers, which live close to the bottom.*

Sturgeons. *They have a primitive heterocercal tail with a large upper lobe on their caudal fin.*

Tuna. *The moon-shaped tail of the tuna creates little drag in water.*

Blackfish. *In sudden bursts, its rounded caudal fin creates cavitation and a sonic boom.*

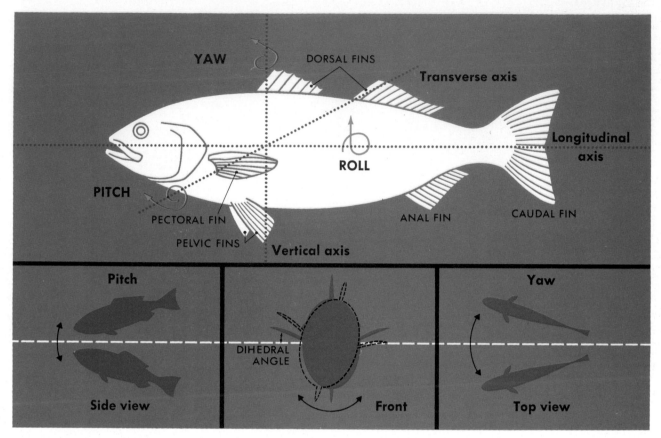

Fins provide stability. Fins of fish resist movement about the animal's body axes, overcoming the tendency to pitch, yaw, or roll. Upswept forefins (dihedral) prevent roll.

Counteracting the Sea's Forces

The world of the fish is three-dimensional with forces pushing and pulling in all directions. These forces are gravity, which tends to pull the fish downward; buoyancy, which tends to hold it up; and drag, which tends to hinder its forward motion. Secondary effects of these forces are pitch, roll, and yaw. In pitching, the fish's head rocks up or down on its transverse axis. To counter pitch, the pectoral fins extend at the proper angle of attack and camber. In rolling, the fish tilts from one side to the other about its longitudinal axis. To resist roll, all the fins are extended. Yaw is the tendency of a fish to turn about its vertical axis. To resist yaw, the unpaired dorsal becomes erect as well as the anal fins. Maneuverability requires the ability to brake rapidly, then all fins and the tail are curved to offer maximum resistance.

Moorish idols. The laterally compressed body of these disk shaped Moorish idols gives them great lateral stability to prevent rolling. And their tail dorsal and anal fins give them even greater ability to control side-to-side yaw. This ability stands them in good stead when they dart among Pacific coral reefs, where the surge of waves could dash them against the jagged coral. The fish's narrowness also allows it to delve into tight little corners of the reefs for food. To maneuver in and out of tight coral crevices, the Moorish idol is capable of turning in less than one half the length of its body.

Fins to Lift, Stabilize, Stop

These requiem sharks, built for speed, have difficulty stopping. Unable to brake themselves effectively, they must make sharp turns. Because of this, few sharks venture in coral reefs where space is restricted, and they roam around the reefs. Their fins are used for maneuvering.

This white-tipped reef shark (above) *glides along on its pectorals using them as if they were wings. The shark's large, white-tipped first dorsal and tiny, dark second dorsal, along with its anal fin, all combine to give it lateral stability.*

A requiem shark (right) *uses its pectorals, twisting one up and the other down, to effect a sharp turn, perhaps to slow down. The combined action of its caudal fin with its pectorals helps make the drastic maneuver successful. The shark's dorsals keep its body from rolling.*

A sea bass hovers quietly by gently sculling while a small cleaning goby picks off parasites.

To Start, Hover, Turn, Stop

The basic simplicity of design of a typical fish or of a marine mammal when compared to that of a lobster, of a giraffe, or of a man, is obviously due to neutral buoyancy and to the absence of strong, large, complex limbs. Thus the muscular structure can be concen-

trated in one solid pack. We have seen that fins other than the tail are used for fine maneuvering which enables the fish to master its liquid environment. By using the proper fins in the proper ways they can stop in midwater or they can hover in one spot, or they can turn quickly and start moving with a sudden burst of speed or at a leisurely pace.

Nassau grouper brakes by fanning its pectorals out to full width and turning them to face forward.

In making these various adjustments they depend largely on their pectoral fins which they coordinate with the movement of their other fins. Sea basses, like most other bony fishes, have swim bladders by which they can regulate their buoyancy. Working together with the fish's fins, this buoyancy control enables them to move about in almost any conceivable manner through their three dimensional world—up or down, forward or backward, left or right.

Many divers in tropical waters have seen groupers, those large members of the sea bass family, hovering patiently near them, sizing them up.

Compressed Bodies

The compressed bodies of yellowtail tangs give them stability by preventing any tendency to roll while enhancing their ability to make sharp turns. They generally move in unison in schools. Even their caudal fins are turned the same way, as they move compactly through the water. Tangs, and their relatives the surgeonfish, swim with beats of their weak tails and by "sculling" with their pectoral fins. Their slender bodies knife

Yellowtail tangs, seen here in a school, offer little resistance to the water they swim through because of their slender shapes, streamlined in horizontal sections.

through the water and offer potential predators a shape difficult to swallow unless it is seized just the right way. Other advantages to the slender shape of the tangs, surgeonfish and other species that have compressed bodies include the ability to slip into narrow crevices in rocks or coral reefs to seek shelter against their enemies.

Living Metronomes

Pufferfish, triggerfish, and ocean sunfish move about in a unique way. Their bodies remain practically rigid and they move by waggling dorsal and anal fins synchronously to the same side, which gives them a readily recognizable appearance. When undisturbed, they leisurely beat these fins left and right like living metronomes.

Though dorsal and anal fins are in charge of

This clown triggerfish, as well as other members of the triggerfish family, are not especially swift swimmers, and may have difficulty outrunning predators. Perhaps to offset their relative slowness, the first spines of their dorsal fins have been modified into a movable unit that acts as a unique defense. The first spine, being longer and thicker than the rest, can be erected by muscles then locked into place by the second. Until this spine is lowered the first cannot be depressed. When raised, the triggerfish's spine makes it difficult for enemies to eat.

most of the triggerfish's routine propulsion, powerful strokes of their tails enable triggerfish to make sudden "spurts."

49

No Lateral Stability

Morays are eellike fish that have no pectoral or pelvic fins. They do have long, low, fleshy dorsal and anal fins, which meet and unite with their caudals. They move by undulating their bodies and those long median fins in a traveling wave. Living as they do in rocky reef areas, they don't need paired fins. Their lack of a high dorsal denies them much lateral stability, and it is not at all unusual to see morays lying on their sides even

Moray eels. Here a moray lies on the bottom bellyside down. They often lie on their sides or upside down.

if nothing is wrong with them. They don't need the lateral stability that large dorsal and anal fins offer, because most of their lives is spent resting on the firm base of rock walls, rather than floating in midwater.

They do not actively seek out trouble, but a diver putting his hand into the hole of one of these creatures could receive a nasty bite.

Dependence on Fins

The pufferfish moves principally by sculling with its pectoral fins. Its small dorsal fin and its caudal fin come into play to some extent, wiggling rather feebly from side to side. When a puffer inflates itself with water, its body becomes even more stiff than it is normally. In such a hopeless situation, it cannot depend on even a slight body motion to help it move, but must rely entirely on its fins.

Because the puffer is so slow moving, it depends on its ability to inflate itself for protection and in some species on sharp-pointed spines that become erect when the fish swells. Its coloration, which tends to blend with the reef environment, also helps it to survive predation. Unfortunately, man can cope with these defenses and puffers and porcupine fish especially are often caught and dried to be sold as curios.

Inflation. This pufferfish inflates with water to protect itself against an approaching predator. That predator may find the inflated puffer too big to swallow or too imposing in its new size.

Two additional defenses puffers have are hardly visible. One is their ability to bury themselves in sandy ocean bottoms by squirming. The other protection is the toxicity of the fish for those eating them. The poison is concentrated in the liver, the viscera, and the gonads, especially around spawning time. Because some species of puffers (the "fugu") are considered a delicacy on the Japanese table despite their poisonous qualities, cooks preparing them for human consumption often must show proof of graduation from a special school that trains them in detoxifying puffers. Seven thousand tons of "fugu" are eaten in Japan each year, but the Emperor is not allowed to enjoy that dangerous delicacy.

51

Large Fish with Little Tail

The ocean sunfish, or *Mola mola,* is big on dorsal and anal fins and very, very small on caudal and pectorals. In fact, the tail fin of the mola is so small that at a quick glance it looks like the fish has none at all. In fact, the mola looks as if it has been cut off just behind its head, which accounts for another of its common names, the headfish. To shift its ponderous body, the mola uses its large dorsal and anal fins, moving them from one side to other at the same time or sometimes to alternate sides. Its tiny pectorals fan out and help a little in steering. Molas are often seen at sea on summer days lolling around on their side near or at the surface. No one knows where they spend most of their lives.

They are improper for human food, but not for sea lions who eventually bite off all their propulsive fins and store them, alive but helpless, on the bottom as food reserve.

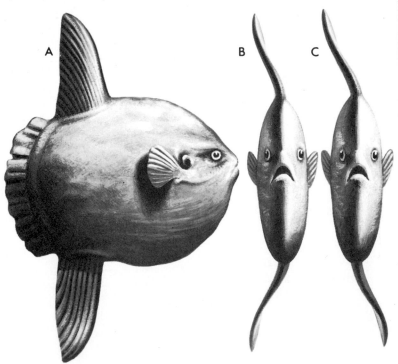

A / A useless tail. Since the tail of the ocean sunfish is so small, it is virtually useless. Instead the fish relies on its large dorsal and anal fins for propulsion. The dorsal fin occasionally sticks out of the water. But when this happens, it ceases to be of much use to the fish.

B / Waving its fins. The dorsal and ventral fins are usually waved back and forth, each traveling to the same side of the fish at the same time.

C / Opposite fin movement. If the dorsal fin is moving in air, the dorsal and ventral fins may get out of synchronization.

Undulating Fins

The seahorse's fins are nearly invisible, but close observation shows that the animal has control of each individual ray. High-speed photography has revealed that each ray is capable of moving at a rate of 70 times a second in an action similar to slats falling in a sagging picket fence. As an undulation passes from one end of the dorsal to another, the seahorse moves forward or backward, or up or down, in its own peculiar, very slow but versatile version of a traveling wave.

Being a poor swimmer, the seahorse usually avoids any areas where strong currents occur. It has sharp eyes and sits with its tail wrapped around seaweeds or gorgonians. When it does choose to swim, it unwraps its tail, straightens itself out, and begins to flutter its dorsal fin. Even so, these slowpokes may take one-and-a-half minutes to cross a one foot area.

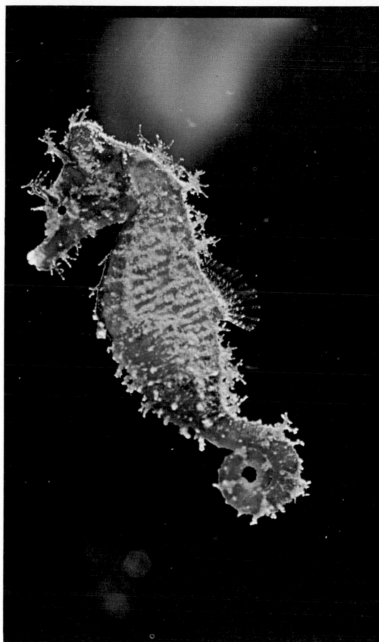

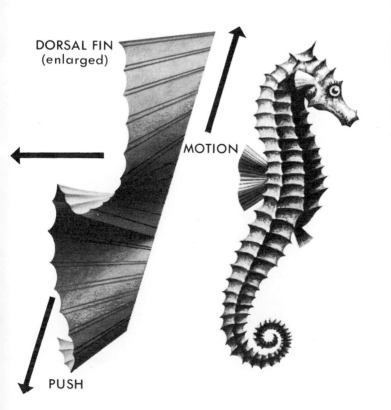

DORSAL FIN
(enlarged)

MOTION

PUSH

Seahorse: Sparse of fins. Looking very much like a chess knight, the seahorse "stands" in the grassy flats that are its usual home. Frequently it grasps a blade of eel grass, manatee grass, or turtle grass with its prehensile tail and remains stationary. But when it does move, perhaps to sip in some tiny zooplankton drifting by, it does so by vibrating its delicate dorsal fin as many as 70 times a second with a wave action that ripples through the fin each time it vibrates. Seahorses have no pelvic fins, a very small anal fin and no true caudal. Only a pair of pectorals, which help somewhat, and the transparent dorsal, which does most of the work, move the strangely shaped, vertically oriented seahorse about.

53

Chapter V. The Jet Set

When a child blows air into a rubber balloon and lets it go when it is well inflated, it will fly about the room erratically as long as the pressure of the air inside the balloon is higher than that of the surrounding environment. This is because a certain mass of air streams out through the narrow neck of the balloon with a force proportional to the dif-

> "Fish that normally swim by more ordinary methods sometimes use a jet-assisted take-off."

ference in pressure, and according to Newton's third law, for every action, there is an equal opposite reaction. It is the reaction that propels the balloon as air is ejected through the rear neck.

The same experiment can be made in a bathtub with a hand-held shower nozzle. When the water is turned on, the shower head is pushed in the direction opposite to the flow of water. If the shower head is raised above the surface, the water jet is expelled into the lighter air, and there is a noticeable increase in thrust. Such thrust would even be slightly bigger if the water jet was expelled in the vacuum of outer space, which indicates that the efficiency of jet propulsion is handicapped in a dense medium. In the sea, jet propelled creatures are capable of swift rushes or very slow cruising, but could not perform rapid, long range migrations like tail-propelled animals.

In the world of the sea there are animals that move about in much the same way as the balloon, making use of jet propulsion. The simplest are "salpa" and jellyfish. Some of the bivalves, or two-shelled molluscs accomplish jetting by very sudden contrac-

tions of their muscles. When the shells clamp shut, water is forced out and the animal travels in the opposite direction of the stream.

Jet propulsion is used by cephalopod molluscs like the octopus and squid. These animals energetically contract their mantle, forcing a narrow stream of high pressure water out through a flexible siphon. The flow of the stream can be directed by the siphon to steer the creature with precision.

Fish that normally swim by ordinary methods sometimes use what is called in aeronautics "a jet-assisted takeoff." This is accomplished by the forceful opening and closing of their gill covers. As the covers snap shut, water is squeezed out and the fish gets an additional thrust forward.

Water-jet propulsion is now used by engineers to move boats in shallow water where propellers could get damaged, as well as for "bow thrusters," transversal jets located across the bow to increase dramatically the maneuverability of service ships or harbor tugs. Water-jet propulsion was also chosen for the first exploration submarine called "diving saucer" in spite of its low efficiency, because it offered an unmatched potential for maneuverability.

This medusa, a jellyfish, is a free-swimming form. It jets by alternately contracting and relaxing its bell-shaped umbrella. It is one of the cnidarians some of which are jet-setters. Medusas are not very strong swimmers; they usually float along with the currents and the winds. But when they do move, it is with jerky, uneven motions. By contracting their umbrella, they force the water under it out the bottom. Normally this drives them upward. But they can also descend and move laterally. When they relax, the umbrella opens and admits a great volume of water, which is forced out with the next contraction. A primitive and inefficient way of moving, but it is sufficient for the jellyfish's simple needs.

Escape by Dance

Scallops usually lie about on the sea floor in great numbers, quietly filtering from the water the tiny organisms that make up their diet. There they live, eating and growing and moving about only rarely. As protection against possible predation, they each prepare for themselves a small depression in the substrate on which they rest. Commercial fishermen seeking scallops locate these beds and drag special nets along the bottom to catch them. Even this sort of threat doesn't arouse the scallop to very much activity beyond feeding and pumping.

If one of the species of sea star that especially prey on scallops shows up in the vicinity of a scallop bed, it turns into bedlam. The scallops can sense their presence by

Scallops. *These many-eyed, many-tentacled animals move by clamping their valves together, which causes a stream of water to shoot out and move them forward.*

chemical clues the sea stars inadvertently leave in the water. And these chemical clues are carried through the water to the scallops. The shellfish react by dancing away in a quick burst of jet propulsion which more often than not leaves the predator some distance behind in a cloud of silt.

If one of the many species of sea stars that do not prey on scallops happens by, there is little or no reaction from the scallops. The chemical messages they send are not feared by the scallops. To escape from the predatory varieties the scallops clap their shells together suddenly by sharply contracting their adductor muscles. Water is forced out from between the shells in twin jets.

Primitive Jet Propulsion

Salps are simply structured animals with a complex way of life, which has an important influence on how they move. Related to the sea squirts, salps are among the most primitive of animals with a notochord, a sort of precursor of the spinal column in vertebrate animals. Many of the salps live not as colonial animals but in aggregates of many individuals having built great chains through the budding method of reproduction. Many of them are bioluminescent—giving off a greenish-blue glow. Each has an incurrent siphon to take water into itself; and each has an excurrent siphon to expel that water. Salps also have bands of muscles, which show plainly through their transparent exterior. They take water in and expel it by alternately contracting these bands of muscles. When they force the water out, they are forced in the opposite direction from that of the expelled water. Jet propulsion! This method doesn't enable them to race through the sea, but it helps them steer themselves a bit as they drift largely at the mercy of tides and currents.

Some of these aggregates of salps contain thousands of animals, each one up to six inches in length and perhaps two to three inches in diameter. Divers in Australian waters have found a horde of salps measuring more than 100 feet in length and nearly 30 feet across. The divers were able to swim around the mass of animals and observe them closely. They found the commune apparently moving along by jetting in their primitive way in the same direction.

Salps. These creatures have almost as little substance as the water around them. They propel themselves by muscular expansions and contractions, and frequently live and travel in colonies, giving them the added strength of numbers.

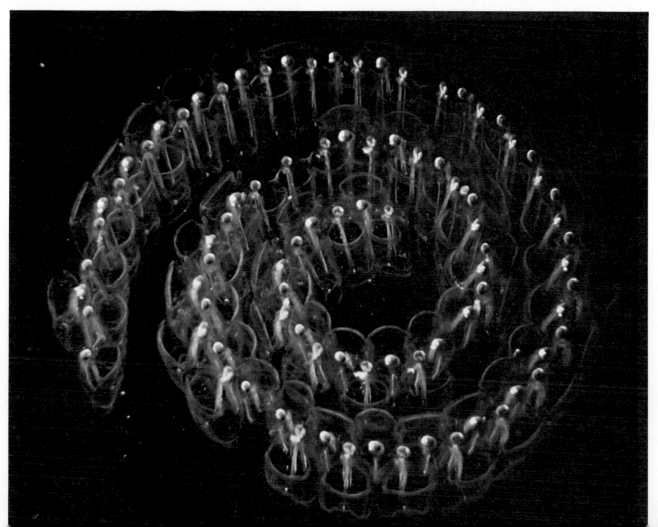

The Jetting Activity of the Octopus

While pursuing the octopus our divers have noted a characteristic escape response. Initially when approached the octopus will freeze and camouflage itself perfectly. It does this by resting on the bottom in a compact mass and assuming the color of the background. As the diver closes in to inspect more closely, the octopus is usually induced to make a run for it. It is at this time that the octopus swims its fastest, using its jet propulsion. Often this movement is accompanied by the release of a puff of ink at each contraction of the mantle. This effort propels the octopus beyond the reach of the diver, but as the diver follows, the octopus

> "The change in tactics may be the result of swimming fatigue."

usually decides to try the camouflage technique again. This change in tactics may be the result of swimming fatigue—we have often noticed that each subsequent flight is a little shorter than the last.

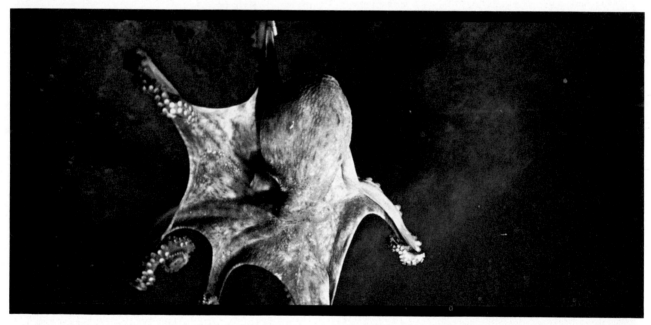

1 / Orienting itself. *An octopus, above, looks around before it descends to the ocean's floor.*

2 / Jetting. *Then, below, it contracts its mantle and expels the water through its siphon.*

3 / To jet again. *Above, it relaxes its mantle and takes in water, then jets the water out again.*

4 / A gentle jet. *The octopus below now jets more gently along the sea floor, seeking a place to hide.*

5 / Stopping. *Below, the octopus extends its arms, stops jetting, and rests on the bottom.*

A Man-Made Jet-Setter

The diving saucer from the research ship *Calypso* is a member of the jet set. It uses the same type of propulsion utilized by a number of animals. Like the siphon of the octopus, the saucer's nozzles can be pointed in any direction. Jets provide not only propulsion for this unique undersea exploration craft but unique maneuverability and safety, as jet nozzles will not as easily get fouled as propellers.

Versatility. *On the bottom the "minisub" above can change direction instantly as it responds to its jet propulsion system.*

Water jets. *The diving saucer (right) blows water from its propulsion jets as it surfaces; this enables its mother ship to locate it more rapidly.*

60

Chapter VI. Walkers and Crawlers

It is thought that the coelacanth lives on the sea bottom in moderate depths of 500 to 1,500 feet, stalking its prey on its four leg-like fins.

Some marine mammals—sea lions, walruses, and fur seals—also walk, or perhaps gallop is a more apt word, at the edge of the sea. Their limbs are now better adapted for movement through water than for movement over land.

Some animals have several means of locomotion. The octopus, for example, employs jet propulsion, but it can also creep along, using its tentacles with their suction disks. It uses the second method in situations where it doesn't have to move quickly—for example when it is bringing building material to its home or decorating items for its garden.

The tiniest of animals, the single-celled protozoans, move by any of three methods. Some have minute cilia, or hairlike projections, they use to set up currents that help them along. Others have whiplike appendages with which they flail their way more efficiently. And some move by extending pseudopods, or false feet, ahead of them, then flowing in after the pseudopod.

Among the snails the conch is a leaper. Because it has a narrow muscular foot, it can't just ooze along like so many other gastropods do. Instead it thrusts its foot into the sand to pushes itself forward. To avoid predators, it can push off and leap perhaps half the length of its shell with every move. Many bivalves, the group that includes clams, oysters, scallops, and mussels, extend a muscular foot and by alternate contractions and expansions creep along.

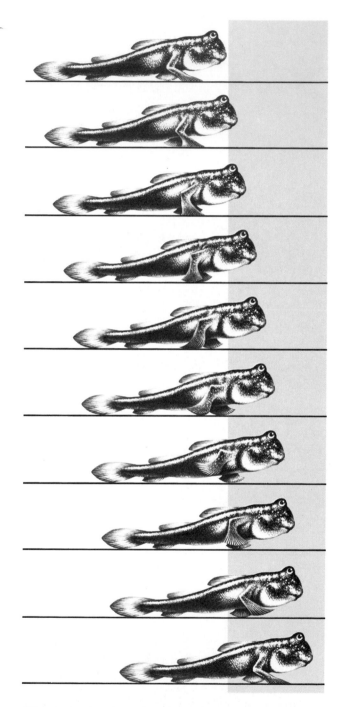

The fish that walks on land. Above you get an idea of how the mudskipper of the Indian and Pacific Oceans cavorts across muddy tidal flats. Out of the water, it raises itself up on its forelimbs and levers itself forward. Then it moves its pelvic fins up and pushes off. Fearless but not foolhardy, mudskippers constantly check their surroundings for danger.

The macruran shrimp walks and swims about in tide pools at the edge of the sea. As it feeds off carrion in the shallows, it crawls about, using its walking legs and its swimmerets.

Cautious Stalking

Lobsters, crabs, crayfish and shrimps are animals that wear their skeletons on the outside. Many appendages protrude from these suits of armor, most of them adapted for special activities. Some are used to hold eggs near the animal's body until they are hatched. Others are formed into claws used for fighting, or for capturing and handling food. Four of their five pairs of jointed legs are used to walk about on the ocean floor.

Their armored suits make walking an awkward process for these many-legged crustaceans, for they limit the distance and

The spiny lobster walks on legs, using swimmerets to help out. These animals spend much of their time in rocky overhangs but when they come out to feed, they stalk proudly along the sea floor.

direction their legs can spread. Because many of these crustaceans are scavengers this method of propulsion is adequate for them to amble across the bottom picking up animal and vegetable detritus. If disturbed, shrimp and lobsters can swim very quickly backward, by flipping the tail underneath in a forward direction.

Members of this ten-legged order (Decapoda) are able to regenerate limbs they have lost. In experiments all ten legs were removed from spiny lobsters. While they regenerated their missing parts, the animals showed a marked craving for calcium carbonate, becoming cannibals if deprived of it.

Walking on Fins . . .

The sea robin is a fish that "walks" on the bottom of the ocean. It walks on its "front legs," you might say. Actually it uses the three lowest spiny rays of its pectoral fins as if they were legs. The way the sea robin moves his pectoral rays might remind an observer of someone drumming his fingers on a tabletop. The rest of the sea robin's huge, fan-like pectoral fins are spread out parallel to the bottom. It also swings the hindmost part of its body to propel itself forward. As the sea robin strolls along, it frequently stops to dig into the sand with its finger-like pectoral spines to root out the smaller animals it feeds on. When it isn't traveling on the bottom, it may swim free in midwater with its pectoral fins tucked tightly in at its sides and overlapping its anal fin. When threatened it may dig into sand.

The sea robin shown here uses three of its pectoral spiny rays and its body to move about on the sea floor.

When sea robins use their pectoral spiny rays to grub in mud, sand or weeds in search of the small animals they feed on, their usual prey are sedentary animals: tiny crustaceans, molluscs and worms. Occasionally they eat herring, menhaden and small winter flounder which requires them to move quickly. Anglers have taken sea robins while trolling for mackerel with spinners, indicating they can swim moderately fast when they are not 'walking' the bottom. Although they apparently do not breed in the northern waters they are not uncommon in those waters, indicating they probably wander there in search of food. So despite their apparent bentic life style, there is evidence sea robins do a fair amount of near-surface swimming.

. . . And Again

The batfish is another, even more grotesque fish that "walks" the ocean floor. It has no swim bladder and therefore it tends to stay at the bottom of the sea, in shallows and in deeper water, where it elbows its way. Its pectoral fins are jointed and heavily built. It is on those strong, jointed pectorals that the batfish walks. It uses pelvic fins to a lesser extent in walking. Sometimes it breaks into different gaits—hopping like a rabbit or loping along in the fashion of wild dogs but not nearly as fast. When batfish do get off the bottom and swim, it's with an ungainly, awkward swimming style. They are broad and flat, and they live in even areas that are free of obstructions. Quite often, as a result of their preferred habitat, they are taken by commercial fishermen in their bottom-scraping trawl-nets. Their flesh is tender and tasty, but batfish have only a small com-mercial value probably because of their ugly appearance. They can be dried out without decomposing and in the Far East dried specimens are hollowed out. Pebbles are put inside, spines smoothed down and the dried batfish serves as a baby's rattle. They are of interest to the aquarium keeper.

The batfishes number about 60 species found almost exclusively in tropical oceans of the world. They grow to a maximum size of about 12 inches and they eat a wide variety of animals, including crabs, smaller fishes, worms and molluscs, which they capture after lying in wait, hidden by their cryptic coloration and sometimes by covering themselves with sand. A batfish, approached by a diver, will 'freeze' or perhaps cover itself with sand, then lie motionless until the diver leaves.

This batfish is awkwardly walking on its elbows along the ocean floor. The elbows actually are a part of the fish's pectoral fins.

Versatility of Movement

A key word for survival in the sea is versatility. This is especially true about the means animals in the sea use for propulsion. Those that are versatile in their modes of movement perhaps have a slightly better chance of survival. By being able to move about in more than one way, animals become a little less predictable. And by being less predictable they increase their chances of escaping predators. Snails, with ponderous shells, generally move quite slowly, creeping forward on a foot that lays down a carpet of slime. They leave a trail that can be followed by a

"Some snails somersault their way to safety, thrusting their bodies forward in jumps as long or longer than their shells."

Nudibranchs (above) are swimmers some of the time. But they also resort to the slower means of locomotion of crawling, especially when they are feeding. Then they move by the muscular wave action like snails and many other bottom-dwelling marine animals.

Whelk. A big muscular foot extends from the shell of the whelk, the large snail pictured at right. And on that foot the snail travels, leaving a trail of slime to grease the material under itself. This whelk carries some unusual passengers—sea cucumbers, which are related to the seastars.

predator with a sense of smell. But when some species of snails perceive danger they can leap into action to escape. Some somersault their way to safety, thrusting their bodies forward in jumps as long or longer than their shells. Similarly, some clams inch along at an amazingly slow rate until alarmed. Then, they can burrow faster than a clam digger can dig. Worms that crawl slowly on the ocean floor much of the time

can take off and swim very much like small sea snakes at a much greater speed. Some nudibranchs are able to swim by graceful undulations of their bodies that work like all other traveling waves. Much of the time they crawl slowly on the bottom and over obstacles in their constant search for food. There are some other molluscs with winglike extensions of the lateral foot that are able to swim with a flapping motion of these extensions. Sometimes, the mollusc is the pursuer and can swim to catch its prey.

Sea cucumbers are perhaps a little slower even than their relatives, the poky sea stars. Some of them have tube feet similar to those of the sea stars, others without tube feet, move by contracting their bodies and by using their anchorlike tentacles to give them traction on the sea floor. Mostly, they move about in their search for food rather than as a means of escape from predators because there are few animals other than man and sea gulls that prey on them. Their tough exteriors help protect them.

How Various Starfish Move

Sea stars have a very complex, not very efficient, system of propulsion. To understand it, we must look at their anatomy and the equipment they have for moving.

These marine animals, often called starfish, have flattened bodies and the most common of them have five arms extending from a central disc. There are some species that have only four arms, others have as many as forty. These are not appendages to the body, but part of it; each arm contains branches of the animal's systems. An arm separated from the central disc can regenerate a new body, and a replacement for the arm will be regenerated by the old disc.

The central disc and arms are covered with a skeleton of shell-like plates or rods that are loosely meshed together to allow the ani-

mal great flexibility. The sea star's mouth is on the side of the body facing the sea floor. On this same side, along the center of each arm is a V-shaped furrow, called the ambulacral groove. This groove holds nerves, blood vessels and a water canal, all radiating from the central disc. Outlining the ambulacral groove are rows of tiny muscular tube feet, which end in suction cups in some stars, points in others. These are protected by movable spines.

It is on these thousands of tubes that the sea star moves. But it is not as simple as extending them in the desired direction and then pulling the body forward. The terrain the sea star moves across is often soft, slippery, uneven or unstable. Unless the star is climbing, say on an aquarium's side, it does not use its suction cups to pull itself forward. Instead, it depends upon the pushing action of its tube feet, and it uses hydraulic pressure

places them firmly, and uses them as levers to push its body along. Some of the feet will extend farther than others to conform to the substrate being traversed. While some are expelling water, others will be filling. All of the sea star's thousands of feet (some are estimated to have up to 40,000 of them) must be coordinated in order to move the animal effectively. It is not surprising that most sea stars are unable to move rapidly. Their average speed is about six inches a minute. A sea star that has been upset is able to right itself in one of two ways. It may pull all of its arms together around the mouth to make its body a tulip shape. Or the star bends its arms away from the mouth until it stands on their tips and topples with some of its feet down.

Having no head or tail, a sea star never needs to turn around. Rudimentary "eyes" and sensitive feelers at the ends of each arm tell the animal the direction it should go. Occasionally the directions can be at variance with one another. One five-armed star was seen attempting to go in all directions. It succeeded in tearing itself into five parts.

Turning over. In this sequence, reading from left to right, an Oreaster sea star rights itself by the so-called somersault method. It uses its tube feet, which are extended, and the suction cups on the ends of them to pull itself around and over.

Basket stars (right) spend the daylight hours curled up into an unrecognizable ball. As nighttime approaches, they unravel themselves, revealing five arms, each with numerous branches that they stretch across currents as a large trap for small prey.

for this thrust. Water that is channeled from the central disc flows into little bulbs above each individual tube foot. By muscular contraction of the bulb, water is forced into the tube through a non-return valve, and extends the tube foot. The tube is shortened by expelling the water through the pressure of the tube's longitudinal muscles and relaxation of those of the bulb.

When a sea star wishes to move, it extends some of its tube feet in the desired direction,

Sea urchins, armed with sharp spines, frequently graze on algae as they proceed along the sea floor at a barely discernible speed. It requires time-lapse photography to see the patterns of their movements.

Slow Movers

Sea urchins and sea cucumbers are relatives of the sea stars and generally move as slowly. Sea urchins have tube feet as sea stars do, only those of sea urchins are more slender and longer, reaching out beyond their spines.

They move by means of their tube feet or their spines or a combination of both. The oral surface, that is, the side that rests on the sea floor, has the tube feet and spines that are used to move about. The tube feet, like those of some sea stars have powerful suction discs at their ends enabling sea urchins

This sea cucumber appears to be lying quietly on a sandy bottom. A few sea cucumbers just don't bother to move around in search of food—they simply let ocean currents carry it to them.

to climb slick vertical surfaces. Using their tube feet, some sea urchins have been clocked at speeds averaging about an inch a minute or five feet an hour. Others have been clocked as fast as six inches a minute. Traveling on the tips of their spines, one species of sea urchins has been clocked at speeds up to five feet a minute—or 300 feet an hour. This species travels in groups in what appear to be extensive migrations. Some sea cucumbers have tube feet which they use for locomotion. These and others without tube feet also use worm-like contractions of their body muscles to move.

Trails On the Seafloor

When man first lowered cameras into the depths of the sea and took pictures of the abyssal ocean floor he found evidence that someone or something had been there before. All over the seafloor were markings of some kind which proved, on close examination of the photographs, to be animal tracks. More cameras with lights were sent down and more pictures were taken and more tracks were discovered. Eventually, man was able to get down to even the greatest depths personally to inspect this mysterious ocean bottom and study the animal tracks firsthand instead of from photographic evidence. What he discovered deep in the abyssal portions of the sea were not only animal tracks but the animals that made those tracks, and evidence of an intense underground life deep within the bottom sediments. At last there was eyewitness, indisputable evidence that there is, in fact, life in the greatest depths of the world's oceans.

What did the scientists find when they investigated the great depths of the sea? What did they see crawling around on the seafloor making those curious little tracks and trails? They saw a collection of some of the strangest animals man has yet found. And they also found some very ordinary looking creatures that might just as easily have been living on the continental shelf.

Among the curiosities they found the blind tripod fish, *Bathypterois,* standing on the

Animals on ocean floor. *Above we can see short-spined echinoderms called sea biscuits, pushing their way along the soft-bottomed ocean floor. They are related to the nearby sea urchins and sand dollars.*

bottom on a couple of rays from its two pectoral fins and on a third ray from the lower lobe of its caudal fin. There were deep sea codfish, sharks snuggling into hollows on the bottom, and sea urchins laboriously moving over the seafloor. There were sea stars and brittle stars and crinoids. They found crustaceans and several kinds of seaworms.

Chapter VII. Other Ways to Go

The quickest animals that live in the sea get around by swimming, beating their fins, undulating their bodies in a traveling wave, or using jet propulsion. Slower creatures walk or crawl on the bottom. Still others stay in one place, taking sustenance from the passing waters.

For centuries man depended on the power of winds or currents to take his vessels across the seas. Finally, by trial and error, a totally submerged propeller with two or

> "Some dinoflagellates travel 150 feet each day. This sounds like a short distance to us—but the equivalent for man would be a daily 2000-mile underwater journey."

more blades was developed. With refinements, it is the motive mechanism found on most ships today. It is tailored to suit the specific vessel it serves, whether it is an atomic submarine, a passenger liner, an oil tanker, or the family launch.

But this is a costly answer to water travel and gives man little prospect for improvement. We look again to other animals for clues. Dinoflagellates, tiny plant-animals, have two whiplike organs they beat against the water to propel themselves, pushing the water behind them. Some dinoflagellates travel 150 feet each day in vertical migrations. This sounds like a very short distance to us, but it is almost two million times the length of the dinoflagellate. The equivalent for man would be a daily 2,000-mile underwater journey, clearly beyond even our most advanced technical capabilities.

Looking to animals more similar to man, we compare ourselves with other mammals that have returned to the sea and find many similarities in technique. Like an experienced Aqualunger, who uses his arms in swimming mainly to turn, a sea mammal with armlike appendages usually holds them close to its body. The turtle's use of its flippers as paddles for swimming is not unlike the underwater swimmer using his arms to do the breast stroke. Nor is the penguin's use of its wings for underwater flying much different.

We have copied the webbed feet of sea birds and the flippers of seals for swimming. Flippers give us more surface to push against the water, but our legs are peculiarly unsuited for an aquatic life. Many serious swimmers, interested in improving their technique, have attempted to emulate the dolphin's swimming movements. To do this, they keep their legs together in one unit similar to the dolphin's flukes and flex their legs from the hips with only a suggestion of knee action. Swimming this way, a man with only one leg can compete effectively with a man who has two. But we are still left far behind in the wake of the dolphin.

Electric motor submarines. With the advent of the Aqualung people could go underwater free of surface connections. But to move about beneath the surface with a minimum of effort and thus conserve air, divers began designing and building wet submarines. The most primitive were towed by surface craft, defeating the purpose of being free of topside support. A step along the road to independence was the pedal-powered submarine, but this forced the diver to expend energy and thus use up his air supply more quickly. The most advanced of these wet submarines are powered by electric motors much as military submarines. Some carry two Aqualung divers, some carry one as pictured here. In this submarine the electric motor is sealed in the stern; it is connected directly to a reduction gear, also sealed watertight, and that, in turn, is connected to the propeller which provides the needed drive.

Man's Undersea Limitations

Basically the size and weight of a porpoise, a human being also displays a comparable anatomy. But the analogy ends there. The porpoise's physiology allows him to stay underwater without breathing ten times longer than man; to dive five times deeper; to swim at least ten times faster; to see clearly above and below water, while man is practically blind under the sea; and to be practically insensitive to water temperature, while man survives but a few minutes in ice-cold seas. We will see in another volume what technology can do (including surgery) to improve human physiology under the sea. But a still formidable obstacle is self propulsion, for which we are basically crippled. The ocean is vast in three dimensions; and swimming unaided, even the most sophisticated *Homo aquaticus* can only be master of a very small piece of territory.

Homo aquaticus. His lungs will be bypassed and his blood supplied with oxygen from cartridges.

Man's swimming capabilities. Compared to sea animals, a human moves through the water with a very low rate of efficiency. The addition of swimming equipment helps considerably, but it still doesn't greatly increase our speed or efficiency. Without equipment, the fastest way to move is to plane along on the surface. Surfers know this and use a flat board to present a plane gliding surface to the water. (They add a skeg, or keellike device, on the bottom of the board to give some small measure of stability.) We can do away with the board, of course, and body surf, as this swimmer is doing. Or we can flail our arms and legs in more or less regular and rhythmic fashion the way a swimmer does and move through still waters.

Iguana. The marine iguana is a mediocre swimmer, using its body and long tail to undulate through the water. It rarely goes more than 100 feet offshore, however, since it finds seaweed close to shore.

The Waggler

The marine iguana tucks its four legs close to its sides and uses its body to wend its way through the sea. While most lizards are terrestrial, the marine iguana of the Galápagos Islands in the Pacific lives in and at the edge of the ocean. Some of the time it lies quietly in the sun along the rocky shore. Some of the time it swims offshore and dives to the beds of algae on which it feeds. To move through the water, it uses the time-honored method of its relatives: the traveling wave. Through evolution, the tail of the marine iguana has flattened laterally in about the same manner as the tail of the sea snake.

Walrus. This mammoth animal spends most of its life in the sea, and it is an excellent swimmer. A bull walrus can weigh close to 4000 pounds, and its thick layer of blubber protects it from the icy water.

Blubbery Athlete

This huge, ungainly, blubbery animal which often wallows on the ice floes of the Arctic Ocean, becomes a graceful, smooth, efficient swimmer when it slips slickly into the sea. The transformation is astonishing because the contrast is so great. On land the walrus moves so ponderously that an observer feels sorry for it. But the same observer must feel a sense of wonder when he sees how graceful the enormous beast becomes in its own element. The walrus tucks its forelimbs out of the way when it oozes itself into the water and uses its hindlegs—broad, flattened, paddlelike limbs, to propel itself.

Adelie penguin. *This is one of the species of penguins which live on the Antarctic continent. It does not fly, but uses its wings to swim.*

Double-crested cormorant. *Large and powerful, with webbed toes and a hooked beak, this bird is a voracious eater and excellent fisherman.*

Webbed-Foot Swimmers

There are fifteen species of penguins, found in various parts of the Southern Hemisphere, and as far north as the Galapagos Islands. Seven of these species live almost exclusively in the frigid waters of the Antarctic which are rich in their basic food, shrimp named "krill." Penguins spend most of their lives in the sea, they "fly" very swiftly underwater using powerful strokes of their short, smooth wings. Their webbed feet are used mainly as rudders. Some species are capable of cruising great distances at seven knots and to dive as deep as 900 feet. Cormorants also "fly" underwater with their big wings and can cover at least half a mile without surfacing.

Manta ray. *Also known as the "devil ray," these animals can weigh as much as 3000 pounds. The manta ray makes beautiful leaps out of the water.*

Underwater Fliers

The manta ray is another underwater flier. There is a certain similarity in the underwater swimming of birds like penguins and cormorants to that of fish like mantas, skates, and other rays. In each case the forelimbs, analogous with our arms, are used for underwater flight. While birds use their wings, mantas use their pectoral fins which are developed into enormous, flexible, triangular, winglike structures. And by flapping they change the pitch of the wing to get the utmost efficiency during up as well as downbeats. The result is a graceful ballet. The manta's reputation as a beautiful swimmer is beginning to surpass its undeserved reputation as a devilfish.

Paddlers

Man's racing oars have flat blades, slightly curved and of the right length-to-width proportion for top efficiency. They're lightweight to enable the oarsmen to swing them easily to the coxswain's call, sometimes as many as 40 strokes a minute in hard-fought competition. The oarsmen's seats slide back and forth in response to their strokes. The most efficient undersea paddlers are the giant sea turtles. Clumsy, almost helpless when

Sea turtles, like the hawksbill above, are so completely aquatic in life-style that only the females ever leave the water and then only to lay eggs on the beach. Once the eggs hatch, the hatchlings immediately head for the sea.

The fastest man-powered boat. The most efficient man-powered watercraft is very likely this eight-oared racing shell (right). Working as a closely co-ordinated team the oarsmen each swing one oar in unison with all the others.

they have to creep ashore to lay their eggs, they are fast migrators in the sea, can reach peak speeds of ten knots, and are very agile.

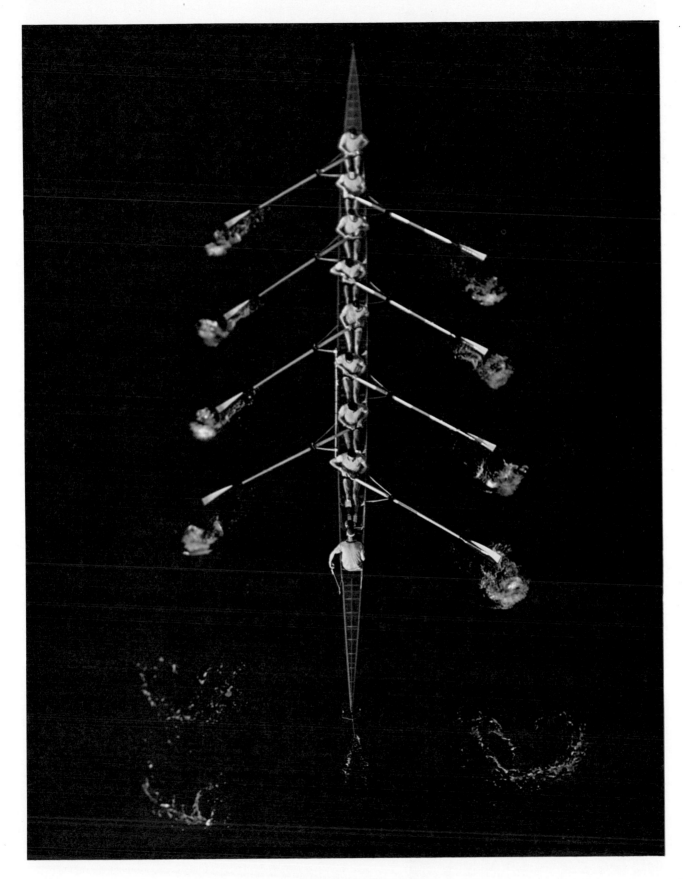

Catching the Wind

By twisting its sail, a tiny velella is able to steer itself slightly, but practically it is at the mercy of the winds and is often stranded on beaches. Beneath its float of air cells, there are many short tentacles with which the by-the-wind sailor stings and captures minute living creatures and passes them to its mouth which is centered on its underside. Great flotillas of these little animals are frequently seen in tropical waters, dotting the ocean for miles in all directions. Fortunately, their sting is harmless to man.

Single triangular sails have plied the coastal waters of the Indian Ocean and the Red Sea since the ancient Egyptian and Phoenician sailors of 5000 years ago. Square sails were common until less than 100 years ago when the more efficient triangular sails became

A living sail. (Above). Usually the tiny iridescent sail of velella, the by-the-wind sailor, stands high in the breeze that pushes the little hydrozoan across the surface of the sea.

Man-made sails. (Right). Like the velella, man has harnessed the wind for many centuries, using it to send him across the surface of the sea in boats like these Arab dhows.

widespread. Until they were supplanted by steam and deisel engines, sails powered all the world's ships. Modern yachtsmen have a background in aero- and hydrodynamics; they are able to tack pretty close to head winds. They have learned also that certain winds can be counted on at certain times of year to help in ocean crossings and competitions. Racing sailboats have become very sophisticated.

Chapter VIII. Hitchhikers

An attractive way of life is that of the marine hitchhiker. It is a rare event to find any of the sea's large and powerful creatures without at least a few freeloaders accompanying it. Whales, dolphins, manta rays,

---

"The hitchhikers' mischief ranges from the harmless capering of the little pilotfish to the deadly attachment of the sea lamprey."

---

turtles, man's vessels, and many others play host to one or several easy riders. The hitchhikers' mischief ranges from the harmless capering of the little pilotfish, swimming effortlessly in the compression wave at the snout of a shark, to the deadly attachment of the sea lamprey, who is not so much interested in transportation as in its host's blood.

There are a number of advantages to be gained by hitching a ride. If one cannot swim, as in the case of the barnacle, or if changing location is difficult, as it is for the sea anemones, the host can take his guest to places where food will be abundant. Species that have a wide distribution throughout the seas have a far better chance of survival than those concentrated in one locale. Then no localized phenomenon—a change in the water's temperature or chemical makeup—will be able to wipe out the whole population.

The distance a hermit crab takes its passenger, the sea anemone, is not the most important factor in this driver-rider relationship. These crabs live in the abandoned shells of other animals and occasionally must relocate to larger ones. The crabs appear to want to carry anemones along as decoration, and they probably gain some

protection from the anemone's stinging tentacles. At any rate, the association has been observed very frequently; the anemone's tentacles are often positioned close to the crab's mouth, thus enabling the rider to catch a few crumbs from the host's table.

Other hitchhikers return no favors to their benefactors. They are along on the ride for what they can get.

Flatworms, roundworms, tapeworms, and threadworms invade the intestines, heart, liver, muscles, and bloodstream of almost all species of fish. In addition to these internal parasites many animals are afflicted with fish lice, a crustacean-like animal that fixes itself with a circular sucker to the outside of a fish and changes its position at will. A slightly different parasite attacks fish in the same way as do fish lice.

Disagreeable as they are, most parasites allow their victims to live. Not so the sea lamprey. With a disklike, rasping mouth, the lamprey attaches itself to its victim, drills a hole in it, and sucks its blood. When there is nothing more to gain from the unfortunate host, the lamprey drops off and sets out to find another ride.

Clusters of whale lice have settled on this right whale's head and body, and they get a free ride wherever the whale carries them. These little shrimplike creatures hook their legs into the skin of the host and gnaw out pits in which they live protected from the wash of water passing over the whale as it swims. The whale louse takes all of its nourishment from its host. Whales that pass through warmer water than does the right whale also play host to a few species of barnacles. These feed on phytoplankton, the microscopic drifting plants of the sea, by kicking it into their mouths with their feet. Attached to a constantly moving whale, the barnacles are carried through unharvested waters, which are often rich in the nutrients that sustain barnacles.

Parasitic Crustacea and Fish

Red wrasse host. Above, playing host to an isopod, this wrasse is deriving no benefit from its hitchhiker. The parasitic isopod is deriving sustenance from its traveling companion.

Isopods are tiny, sometimes microscopic crustaceans that inhabit the waters of the world. There are some that have left their benthic habitat to ride the coastal waters aboard a finned host. Most isopods live on the bottom under rocks or camouflaged among algae and sponges. One group of isopods, commonly known as gribbles, lives in pilings eating wood like a termite and may travel only a few inches in a lifetime. They are all relatives of the common sow or pill bug. The parasitic isopods are travelers, although not by their own means. These are crustaceans which attach themselves tenaciously to the skin of a fish and feed off its life blood. Others grasp the gills or some other part of the luckless fish that must support it. Some species of parasitic crustacea or fish lice, in fact, are choosy about where

they attach themselves to their unwilling hosts, selecting a specific part of the body. To attach themselves in a manner that makes them almost unshakeable, these crustacea use six or seven pair of sharp penetrating hooks.

The shape of such a parasite probably does not affect the swimming efficiency of its host appreciably but its size may. Now and then a young fish is seen with a large isopod attached. Such an isopod may create a wound open to infection. It is the secondary infection of bacteria or fungus that may harm the host more than the actual isopod. When a host dies, the uncaring parasite leaves to find another one to support it.

Remoras and Shark

Remoras or suckerfish are among the strangest hitchhikers in the sea. For they benefit in the commensal relationship they enjoy with other sea animals, and their companion is unaffected. Stranger still is the manner in which a remora attaches itself to its commensal. Atop the head of the remora is an oval shaped suction grip device that looks very much like the grill of an automobile or the grating on an air conditioner. By moving the transverse parts within the oval, the remora can attach itself firmly to any surface. By moving those parts in another way, it can release itself. But this attachment process has no effect on the commensal. It doesn't even leave a mark. The strangest part of all this probably is that the oval shaped organ that the remora attaches with is really its dorsal fin which has migrated forward to a position atop the head and changed in form and function. In other fish the main purpose of the dorsal fin is lateral stability. Not so in the remora's case. It is solely for attaching itself to a commensal. Remoras frequently attach to large animals other than sharks. Groupers—the big 500 or 1000 pounders especially—sea turtles, whales and even, occasionally a human diver have attracted remoras and carried them about. Mainly, the remora gets a free ride out of this relationship. But in addition, it is there when the commensal partner is feeding. If that partner should drop a few scraps of food, the remora is quick to detach, snatch a morsel and reattach. It is also believed to clean its big host of parasites.

Shark with chin fish. *In the photograph below a whale shark is seen with many remoras attached to its chin. This relationship is called commensalism—the remoras are getting a free ride and some food, and the shark is not being harmed.*

Dolphins Riding Bow Wave

As a ship plows its way through the ocean it creates waves. On these waves occasionally, dolphins come to play and ride. Sailors at sea have often seen these playful mammals riding the bow wave of their ships with effortless ease. Dolphins somehow have learned how to get a free ride. From all indications they are not seeking food nor are they being lazy. They probably simply enjoy the game of surfriding.

The theory of how this phenomenon works has been described by biophysicists. As a ship plows along, water piles up in front of the ship's bow and is forced forward. The waves that are thus formed move with the ship. Dolphins typically approach a ship and scout around the edges of this bow wave area, perhaps five to 10 yards from it. This indicates that the dolphins may be "feeling out the pressure field."

Once they have ascertained the conformation of this pressure field, they can either

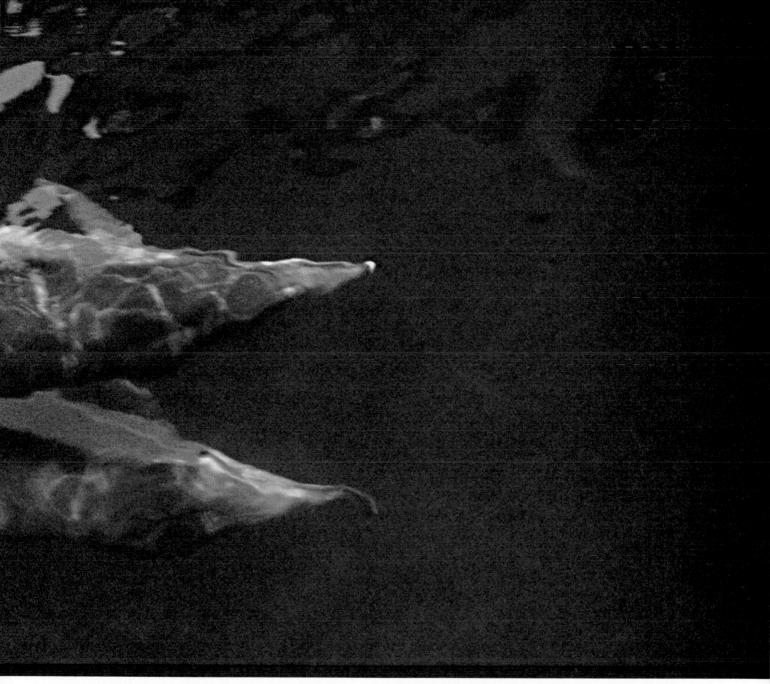

swerve into it or swim away to play else-where. If they do move into the bow wave, they position themselves on its forward slope. Gravity keeps them moving down-ward along the front slope of the wave. And the position in which they hold their flukes gives an additional upward thrust.

As they ride alongside of ships at sea, dol-phins give the appearance of leading the craft toward some safe haven. This impres-

The bow wave of a ship provides a free ride for this pair of Atlantic dolphins. When they are not hitch-hiking, dolphins are capable of out-distancing the fastest ocean liners.

sion is at the origin of the legend of the col-onization of Crete many centuries B.C. Sail-ors have always looked upon dolphins as harbingers of a safe voyage. The very play-fulness of the animals, is another factor that leaves a feeling of good will about dolphins.

Chapter IX. Getting the Hull Out of Water

The sea is a strange environment—sometimes it seems friendly, sometimes not-so-friendly, and sometimes downright malevolent. A denizen of the deep may be wrapped securely in a watery blanket one moment and held in bondage before an onrushing predator the next, unable to flee quickly enough. Different inhabitants of the oceans have adapted to this changeable way of life in different ways. One of the most spectacular methods of escape is to get completely clear of the water even if for just a few seconds. To illustrate—a member of a scientific expedition in the Gulf of Panama was swimming off an island with a sheer rock cliff for a shoreline, when he was approached by a shark. As the shark angled toward him, perhaps only out of curiosity but perhaps with mayhem in mind, the diving scientist, unused to being explored by sharks, tried to scale the sheer rock cliff. He cleared the water completely before flopping back in with a loud splash. As he looked quickly about him, he saw the shark was no longer in sight, and he was able to get back to the boat from which he had been working. Even that brief absence of a second or two from the water was enough to send the prowling shark off looking for something else.

Survival isn't the only motive for getting clear of the water. Some animals quite literally jump for joy, frolicking in the briny is a pleasurable pastime for them. Some other sea creatures leap in anger, fear, or hunger. And there are a myriad of mysterious reasons for jumping, reasons scientists still seek. Whatever it is, the ability to get clear requires special mechanisms as well as special motives. To break free of the sea, a marine animal, whatever its weight, must reach a certain vertical speed. For example, to leap two yards high above the surface, a creature must break the surface with a vertical speed of twelve knots, whether it is a sardine or a blue whale. To reach eight yards, the speed must be twenty-four knots! Consequently, only fast swimmers, large or small, have the privilege of jumping into the atmosphere.

> "Some animals quite literally jump for joy. Frolicking in the briny is a pleasurable pastime for them."

The ponderous cowfish is hardly likely to leap clear of the water even with a healthy riptide behind it. It is just not adequate for a venture into the air. So equipment, favorable conditions, and motive are all factors to be combined if a member of the undersea world is ever to make brief intrusions out of its element.

For a variety of reasons men have also sought to get their bodies and their vessels out of the water. Some have acted to escape danger; some to express aggression; some to develop improved transportation; and some just to have fun. Engineers have designed all manner of devices to get us up and out of, as well as down and in, the water.

A pair of mute swans, after taxiing down the watery runway of their home pond, lift off and get airborne. They use a tremendous amount of energy and must paddle along the surface with their large, webbed feet before they can break free of the water to fly at speeds up to 45 miles an hour. Their speed through the air is far greater than their speed swimming through the water, although their feet are well adapted for surface swimming. Being able to fly gives the swans greater range than if they were restricted to their graceful but slower swimming or their awkward waddling gait over land. Some ducks, with a tremendous burst of energy, can leap directly into the air from a floating start in the water. But most ducks, geese, swans, and other water birds must paddle along the surface to become airborne.

Fish That Fly

The adaptation that has enabled the flying fish to survive through millions of years of predation is its ability to get up and out of the water for extended periods of time—long enough to escape predators. In this case an extended period of time may last up to 20 seconds. As this adaptation continues to develop, it may someday in the far-distant future enable flying fish to soar through the air above the water for minutes at a time. One flying fish was seen to reach a height of 36 feet above the sea's surface. And 20 foot high flights are not uncommon. Soaring along at speeds up to 35 miles an hour, flying fish may travel 250 yards or more in a single glide. However, flights of 60 to 70 yards, lasting six seconds, are more typical, but can be repeated several times in a row with a short dip of the tail.

Typically, a flying fish launches itself through the surface first by gathering speed swimming upward toward the surface. This speed helps it leave the water. With the enlarged lower lobe of its caudal fin trailing in the water, it moves it rapidly from side to side and extends its broad winglike pectoral fins. The forward thrust given by the caudal fin plus the lift provided by air acting on the pectorals help the fish become airborne. When it drops back to the surface, it is the whirring action of the caudal fin in the water that helps the fish repeat its flight. Often the predator follows the flying fish, seeing its silhouette through the surface, waiting for it to splash back. In this case, the flying fish tries to fool its chaser by making a right angle turn in flight. Some fish, notably the dorado, have been seen to chase flying fish right out of the water, sometimes capturing them in their mouths in midair.

A take-off run is being made by this flying fish, much as man-made aircraft do. It starts building up speed as it flashes upward toward the sea's surface.

Clearing the water. The lower lobe of its caudal fin remains in the water for a moment, whirring like a propeller.

Emerging, its extended pectoral fins act like wings do on an airplane and give a certain amount of lift. Its pelvic fins help a little too.

Soaring. As the fish gains speed it expands the pelvic fins, providing lift to raise the tail above the water. It then soars off.

Difficult Journey

After they have spent three or four years in the open ocean, feeding and gaining power, the Pacific salmon undertake their final journey to the stream where they were born. Obeying a mysterious instinct, these magnificent swimmers let nothing deter them from their destination—their mating grounds. There salmon eggs will have a good chance to hatch, and the fry will have abundant food to sustain them until they are large enough to tackle the trip seaward. The returning adults travel through rushing rivers at a rate of about five miles an hour.

Facing turbulent waters. The general flow of water in a river is downstream, but when water tumbles from the top of a fall, eddies develop in the pool below. In these turbulent areas the direction of the flow is constantly changing, sometimes surging toward the surface and sometimes even back upstream. If the salmon can get into one of these eddies, it can take advantage of the backwash there to give it the boost it needs for spectacular leaps.

Rapids and falls are encountered frequently. Then the salmon must leap—spectacularly—to heights of 8 to 10 feet. For some salmon the challenge is too much, they literally batter themselves to death. We know that to leap a waterfall 10 feet high, the salmon must have reached a *vertical* speed of fifteen knots. As the rivers are generally very shallow they can only meet this challenge by speeding horizontally, just below the surface, and suddenly steering their momentum upward by a tremendous effort of their pectoral fins, or by rolling on one side just before jumping, bending sharply and forcing themselves in the air by a last stroke.

Calmer waters. But often the downstream flow of water over a fall is not rapid, or the fall is not vertical. In these situations there is little turbulence at the bottom of the fall to assist the salmon, so the powerful swimmers simply swim up the fall. When the fish reaches the top, its enthusiastic swimming sometimes carries it right out of the water. Then, after a momentary tailstand, it plops back into the water and continues its arduous journey.

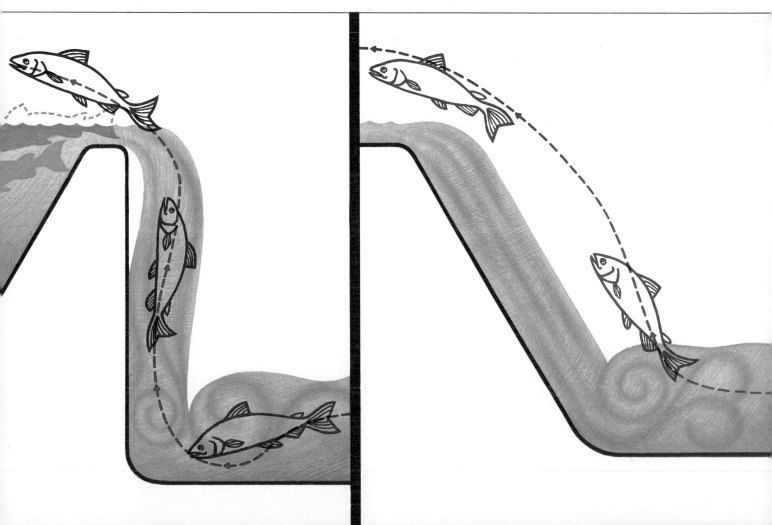

Jumping Billfish

Swordfish, marlin, and sailfish, collectively known as billfish, are known as "game" fish. When it has been hooked by a so-called "sports" angler, the billfish leaps spectacularly out of the water as it attempts to shake loose the barbed and sharpened hook that pierces its mouth. It is no game for the game fish. The angler must keep a taut line to prevent the fish from escaping its intended fate as a futile trophy and photographic model at dockside.

Sailfish, with their high dorsal fins, and marlin, which share a lunate tail with their fellow billfish, are fast, powerful, and they range the open seas of the world. Both have a reinforcing ridge, called a careen, on their caudal peduncle, the narrow part of the body at the base of the tail, that helps streamline transversally the inefficient part of the tail. But the main source of thrust for the phenomenal leaps made by these great fish comes from the segments of powerful muscles in the rear one-third of their bodies: marlins angle sharply upward from beneath the sea's surface and push themselves upward out of the water, seemingly dancing on their tails as they shake their heads from side to side violently in an effort to dislodge the hooks from their mouths. When the fish's energy is virtually all spent, anglers can bring them alongside their power boats to hoist them aboard.

Out of the water. Above, a marlin, sometimes weighing close to 1500 pounds, jumps with the greatest of ease when trying to dislodge a hook.

Sailfish close-up. This model (at right) depicts the wonderful leaps of a sailfish in the open sea. Its long bill not only is an effective weapon but may ease water-resistance.

Leaping for Survival

Mullet often leap clear of the water as they seek to escape danger. The danger from predators may be caused by men seining, as in this picture, or by hungry dolphins or fish. By getting out of the water for even a brief second or two, the mullet may be able to elude the peril that stalks it. Often mullet may be seen leaping by the hundreds from a seemingly placid sea. But somewhere beneath them some danger lurks, probably

Mullet. These leaping fish appear to have boundless energy. They jump to escape being caught by man and their fellow sea-dwellers.

already making a dash into the midst of the school and decimating its numbers. The position of its pectoral fins—high on the sides of the body just behind its head—helps the mullet fling itself clear of the sea. Schooling also helps the mullet in its fight for survival through safety in numbers. Plus the quickness of its movements.

Leaping for Play

Sea lions cavorting in shallows or playing in deeper seas often leap from the water. Their streamlined shape and muscular bodies with powerful hindlimbs and flippers enable them to do this. Sometimes their leaps are conventional arcs through the air with a headfirst reentry. Other times they may do backflips like this one. Typically, a sea lion will pick up speed underwater and then break through the surface, arching upward and

Sea lions. These mammals are strong swimmers. They move by undulating their whole bodies as they dive deep for food.

forward through the air before reentering the water. They may leap clear of the water when they chase fish as well as when they are courting or amusing themselves. Their playfulness and their sleek appearance make them favorites to people along the California coast where they are frequently seen gamboling off the rocky shoreline.

Whales Clearing the Surface

Whales, like people, often like to vary their pace. Just as people will sometimes take to running or jumping or hopping or skipping, so do whales apparently enjoy lolling about on the surface, jumping high into the air, swimming on one side or the other or even on their backs. Of these variations of movement by whales, jumping, or breaching as it is sometimes called, is the most spectacular. Whales come to the surface in a variety of ways and for at least two reasons. Being air breathers they must come to the surface for air. If they are lazing along at or near the surface, they allow themselves to rise vertically until they break surface. Quickly, they will exhale the breath they have been holding, take in a new breath of air, close their blowholes and submerge quietly. Or, if they have been moving along underwater at a good rate of speed, they surface at an angle —about 30°—exhale, inhale, close off and dive back in, all in the space of a few seconds.

There are other times, though, when whales seem to enjoy leaping high into the air, flopping back into the sea with tremendous resounding smacks against the water's surface.

Humpback whales especially seem to enjoy this sport, slapping the water with their flukes. Sperm whales also jump clear of the water only to fall back in, raising gigantic columns of water. Such spectacular jumps are probably the conclusion of very deep dives, when the whales, out of breath, speed

A gray whale breaches as it checks the Baja California coastline near its breeding ground in Scammon's Lagoon. The bay where gray whales breed is protected water under Mexican law.

up from 3000 feet for air at more than 20 knots and leap.

The gray whales which migrate along the California coast each year engage in something called spyhopping. Observers believe the gray whales spyhop to check their location along the coastline, which might be termed line-of-sight navigation.

Jumping for Air

Dolphins are the high jumpers of the sea. How and why they jump has been the subject of study for years. Scientists are still seeking more knowledge about this interesting behavior of dolphins.

To catapult themselves up and out of the water, they use the mighty thrust provided by their flukes; these are powered, in turn, by masses of lumbar muscles, connected with their tails by a series of tendons. Their rigid vertebral columns and these powerful muscles give the dolphins the strength and the speed needed to launch themselves.

Sometimes dolphins leap high into the air, re-entering the water head first or belly-flopping back. Sometimes they make flat jumps, arcing out of the water and splashing back in again.

There are two main speculations as to why dolphins behave in this manner. First, they

do it for the fun of it, jumping as a form of play. Second, they leap to get air which they need because they are mammals. The second theory is explained by noting that simply rising to the surface for a breath of air without leaping clear of it causes a drag and turbulence and jumping eliminates that additional drag. The first point seems just as valid in view of the astounding jumps for meager rewards that dolphins perform in captivity. Some captive dolphins have been noted jumping to a record of about 22 feet.

Pacific white-sided dolphins. This marvelous scene has been witnessed by many an ocean voyager. Dolphins leap out of the water seemingly for the sport of it—and to the delight of those watching.

Orcas and pilot whales also submit to training for spectacular shows.

To leap to these heights above their pools, dolphins and whales require a takeoff depth of only about 20 feet, which suffices for them to gain the speed of 24 knots required for such performances.

Skimming Above the Water on Underwater Wings

A hydrofoil boat is raised above the surface to escape the resistance of water. These craft start from a stationary position in the water exactly as conventional ships do. At low speeds they remain with their hulls in the water. But as they gather speed, submerged wings called hydrofoils provide lift and raise the hull clear of the water. These act in much the same manner as wings on airplanes, designed to take advantage of the

imbalance of pressures above and below the foil, which at high speeds raises both foil and ship. A submerged propeller provides the ship's power.

The hydrofoils are generally one of two types: those that are fully submerged, and those that pierce the surface. Those fully submerged are very similar to airplane wings, fitted with ailerons that are con-

108

stantly adjusted by autopilot to maintain proper depth in all wave conditions. Surface-piercing foils are attached to the hull by struts that form a V or U and give the ship stability. This provides a very smooth ride in water that is not too choppy. Hydrofoils are used as passenger ferries in many of the world's waterways. They are employed by commuters in water-oriented cities like New York and Vancouver. They provide rapid crossings between Sicily and the Italian mainland, and Florida and the Bahamas.

Popular ride. *One of Sea World/Atlantic Richfield's hydrofoils glides over Mission Bay near the marine life park in San Diego, California, at speeds as fast as 35 miles an hour.*

Despite their advantages over usual surface ships, hydrofoils are still of limited use. Their foils are subject to great stress at high speeds, and their size cannot be expanded without an undue increase in weight, which would require an enormous expansion of the power system. Hydrofoils are not yet economically feasible for general use.

Riding a Cushion of Air

The Hovercraft, also known as the air-cushion vehicle, or ACV, rides above the surface of the sea on a pad of compressed air provided by powerful fans. While the air keeps the craft above the surface of the water, huge aircraft propellers drive it at speeds far greater than those of conventional vessels. By remaining aloft, the ACV is free of resistance from the water that reduces the efficiency of surface vessels and submarines.

Hovercraft ferries cross the English Channel between Dover and Boulogne, Calais or Dieppe on a regular schedule at speeds of up to 50 knots. The crossings of this normally rough body of water are either smoother than the ones made by other ferries, or impossible, if the sea is choppy. Hovercraft are very sensitive to the action of winds, and in storms they drift considerably. When a Hovercraft encounters waves higher than its normal flight altitude (which is proportional to the size of the craft) it must either settle

Hovercraft. This man-made vehicle travels across water just above a cushion of air provided by the downward thrust of its jets.

on the water or ride contour over the waves. Following the first alternative, the Hovercraft is subjected to the same water resistance as a surface vessel, and its speed is greatly reduced. Passengers in a Hovercraft flying contour over high waves are given an amusement park ride.

Hovercraft are coming into increasing use because they are able to function on ice, mud and hard ground as well as over water. As their efficiency increases with size (doub-

ling the size requires doubling the power, but the payload capacity is quadrupled), Hovercraft are expected to be the ocean cargo vessels of the future. With their great speed, they could outrun or circle major storm centers.

Chapter X. Toward Higher Speeds

Man has used the oceans as a transportation route for thousands of years. For efficiency, it is unfortunate: the surface of the sea is the last place to navigate! Hulls are half submerged, half above the surface; and this creates monumental problems, among which is the necessity that the ship be a compromise between hydrodynamics, aerodynamics, and seaworthiness in storms. Practically no fast-moving animals have adapted to stay on the water to cruise. Surface ships will sooner or later be bound to disappear—with the exception of yachts.

In underwater propulsion, to reach higher speeds and better efficiency, the challenge is to move through water without moving water, because moving water requires transferring some precious kinetic energy to the surrounding medium, which amounts to waste. We have seen that to achieve such swift journeys, fish, whales, or man-made submarines streamline their designs, avoid all protuberances, and develop very smooth skins or hulls. But this is only half the problem. Whether we consider a tunafish or a military torpedo, it will only move if thrust is applied, and the bigger the thrust, the higher the speed. To generate thrust, a power plant is needed, along with a power transducer system. Finally, some energy reserve must be stored to meet unpredictable demands.

Power plants are of three main generations: the cold power plant used by most cold-blooded animals, such as reptiles and fish; the warm power plant of birds or mammals, and the lukewarm power plant found in some rapid ocean fish like the yellow-finned tuna. In the cold-blooded system, there is practically no "heat overhead;" the energy needed by an immobile fish is negligible, and

it could stay there for weeks without eating and suffer very little. Most of the food consumed by the conventional fish is used to grow, to reproduce, and to *move*. This seems to be an advantage over diving birds and mammals, like penguins and dolphins, which have to maintain a high central temperature even if they remain at rest, and cannot avoid substantial heat losses. But cold blood is seriously handicapped against warm blood if high speeds have to be maintained: high output power plants, whether man-made or natural, have a power/weight ratio that grows with the temperature of the "heat source;" no cold-blooded fish will ever compete with mammals in transoceanic marathons. The "lukewarm" yellow-finned tuna has developed a "counter current" heat exchanger system between veinous and arterial blood to get some advantages from both systems.

The "power transducer" can be the jet funnel of the octopus, which develops considerable thrust, but has poor efficiency; the tail fin of a swordfish; or the "rotating fin" (the propeller) of a nuclear submarine. At least one-third of the energy is lost at the transducer level of the fast swimming machine.

Jaws had a great influence on higher speeds. The most primitive fish, living over 300 million years ago, were probably very sluggish swimmers. With heavy armored bodies, they slowly meandered along muddy bottoms feeding on whatever organic debris was available. Their circular, jawless mouths were capable only of sucking. Speed was of no use. Eventually some fish developed a primitive movable mouth with jaws allowing them to actively seek out other animals as a source of food. Those fish sought as prey then had to avoid capture by swimming; any stragglers were eaten, only the more rapid swimmers survived. Their success meant that only the quickest, most streamlined predators could survive in competing for food. If it were not for jaws, body shape, and propulsion, fish would bear little resemblance to what we see in these snappers.

Using Bursts of Speed

Bursts of speed are used by slow or sedentary creatures like the triggerfish or the octopus; by transoceanic cruisers like pilot whales; and by swift swimmers like the barracuda. To perform such "rushes," a creature must have the ability to generate a strong acceleration of its body, an instant tremendous thrust forward. The necessary energy must be readily available. It has to be stored, chemically, *inside* the muscles concerned;

the energy conveyed by the blood from the central plant will come later to restore whatever was consumed locally. The translation of the energy into a large, short-duration force is best illustrated by the jet squirts of squids—they use such pulses constantly to prey on flying fish at night or to escape porpoises, and they remain almost motionless between two darts. Another typical short-impulse transducer is the deep, long, soft tail of groupers. When they spring to attack an intruder in their territory, the first stroke of

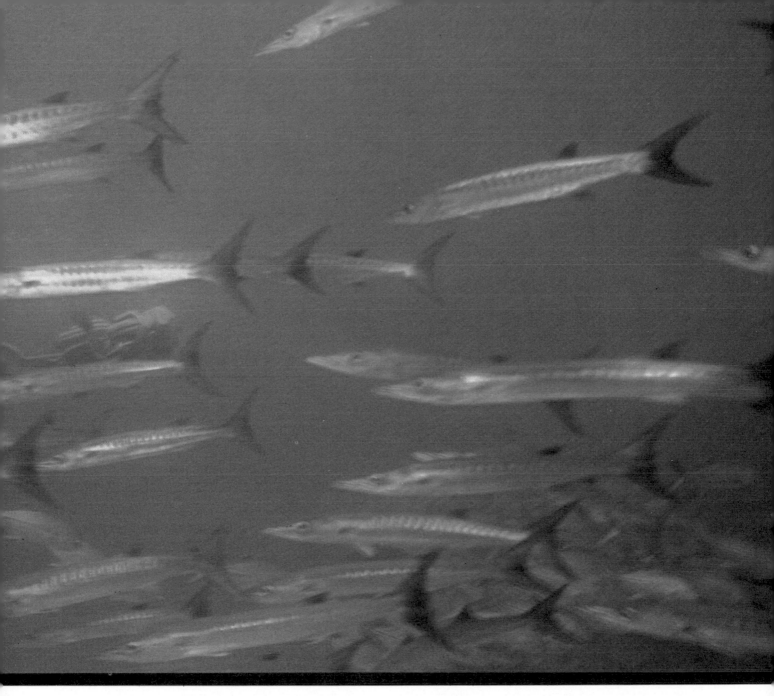

their tail is so powerful that it creates a strong sonic boom due to cavitation. Such organic "explosions" are familiar to divers. Open-ocean fish, like mackerel, tuna, and barracuda that are well streamlined and powerfully equipped to cruise sustain fairly high speeds as a routine, but are also capable of accelerating suddenly for short bursts of speed. These are needed mainly to catch a prey, to escape a more powerful predator, or for the incredible love dances of such strong fish as ocean jacks in the spawning season.

School of barracuda. It is typical of barracuda to remain apparently motionless for a while and then dart quickly off without any apparent motive.

This could not be achieved by those dark muscles that energize the tail at cruising speed, because they receive just the amount of fuel that the central plant can deliver on a regular basis. Other muscles, light in color, are called for, and these have a built-in storage of energy.

Shapes of the Fastest

It has been established earlier that all fast swimming marine animals have the same basic streamlined body profile. To refine streamlining at high speed, the fastest fish have developed grooves or depressions where fins can be neatly tucked away so that nothing protrudes from the living projectile. The flow of water into the mouth and along the gills has also been eased by inner fairing in every detail. Another refinement is the development of laterally streamlined keels on the caudal peduncle: these add the finishing touch to the tuna's tail—a masterpiece of design. Broad and short, highly efficient when the maximum thrust is generated, the tail, in front of the fin, becomes horizontally compressed and tapered on both sides. Such a shape at the same time reinforces the tail, reduces the cross-section of the inefficient part of the tail, and fairs it

so that it knifes alternately with a minimum of water displaced. These caudal keels, shaped in "inverted streams," are found in such animals as marlin and killer whales, which have extremely different stories of evolution.

A / Wahoo. *This member of the mackerel family has all the features of a typical fast fish.*

B / Dolphin. *These members of the order of cetaceans, like their larger relatives, whales and porpoises, are masters in the art of motion.*

C / Greenland shark. *All the pelagic, or open-ocean, sharks tend to be fast-moving to counter their lack of swim bladders.*

D / Blue whale. *The largest creature on earth, this cetacean is not cumbersome but moves with grace and deceptive speed.*

E / Swordfish. *A fine-pointed sword ahead of its body offers some degree of lateral streamlining to this fish.*

F / Tuna. *The dorsal fin on the tuna fits into a slot in its back when the fish speeds off.*

Mackerel are slender, torpedo-shaped fish, which usually travel in fast-moving, well-disciplined schools. They are also fork-tailed.

Jacks have laterally compressed bodies, forked tails with reinforcement at their base, and long, curved pectoral fins.

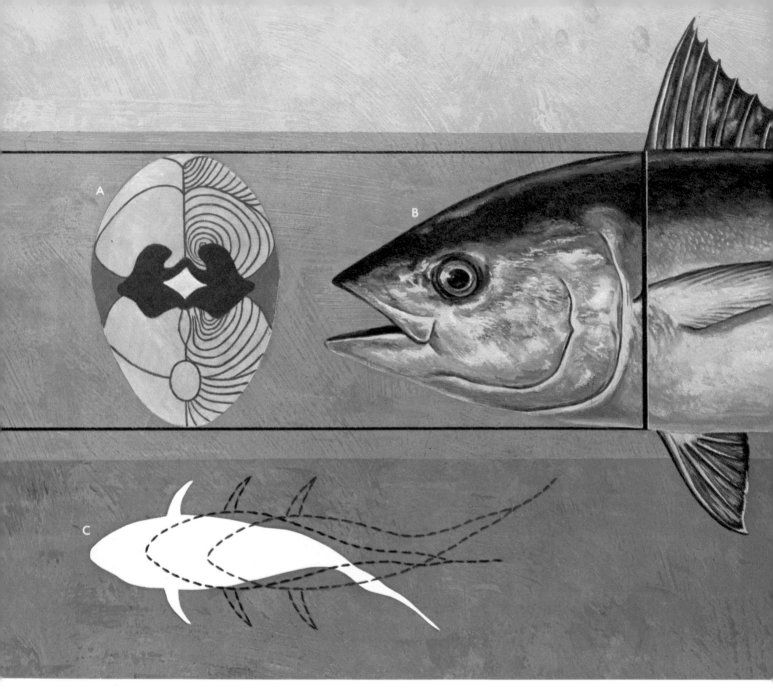

Anatomy of a Tuna

The bluefin tuna is one of the fastest fish because of its design, the high "aspect ratio" of its tail fin, the reinforcement of its caudal peduncle, its powerful muscles, and the chemistry that supplies energy to fuel its fast motion. Laboratory measurements prove that the limiting factor is the amount of oxygen extracted from the water by the gills: per volume, water contains 25 times less oxygen than air. The mouth of the tuna is relatively small to create the least possible drag; the quantity of water it can pump in while swimming, is proportional to speed. This explains why the tuna's metabolism is in equilibrium only at cruising speed. The tuna is sentenced to be a perpetual traveler. It has two different types of muscle tissue which provide a high gear and a low gear into which the tuna can shift.

A / Cross section. The tuna's powerful red tissue and white tissue are clearly shown. Red tissue probably provides the power for slow cruising speeds,

118

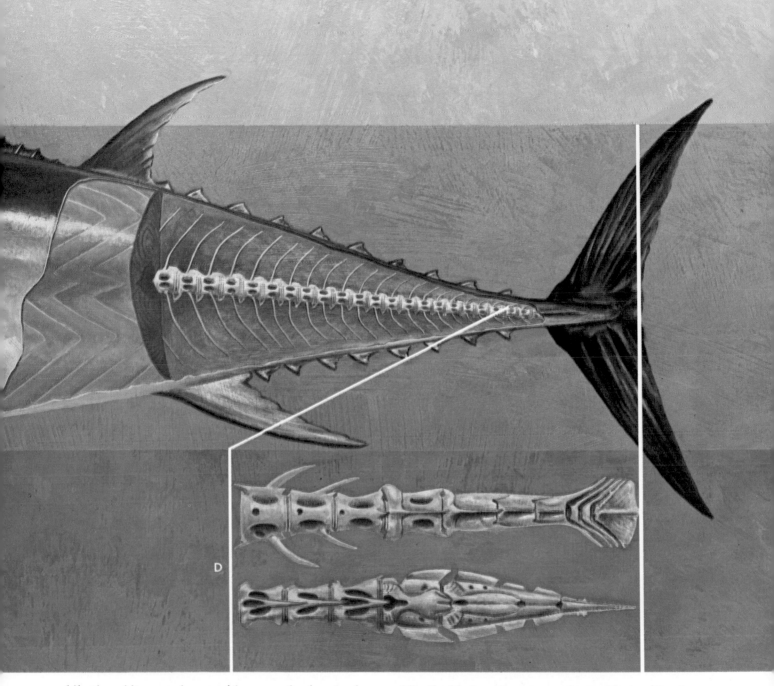

while the white muscles provide power for bursts of great speed. The red tissue derives its color from a blood-like pigment that stores oxygen, providing greater energy. This means higher metabolism, which in turn means higher oxygen demand. To get the additional oxygen it needs, a tuna swims constantly, its mouth open to allow large quantities of water (and thus oxygen) to pass through its gills.

B / Side view. The clusters of muscles, or myomeres, provide the thrust that drives the fish.

C / Powerful muscles. These give a tremendous thrust to the tail and tail stalk, or hind portion, of the body. The front two-thirds of the fish remains rigid

while the hind portion moves from side to side at a rate of about 15 beats per second when the fish is cruising. To achieve this motion, the tuna has two double joints—one at the base of the tail itself and the other at the base of the caudal peduncle.

D / Additional reinforcement of the skeleton. A bony keel provides the leading edge that smoothly separates the water as the tail beats from side to side. Unlike some fish, tunas have tendons that connect the muscle masses with the base of the tail to operate the tail. In the top view the bony ridge that appears externally on the caudal peduncle is plainly visible in the skeleton. The side view shows the ridge in the hindmost part of the backbone just before the tail fin.

119

Tarpon. It is an endangered fish, relentlessly pursued by fishermen. The tarpon nurseries, in mangrove swamps of southern Florida, are destroyed.

The Endangered Silver King

The Atlantic tarpon is a shiny silver fish with extremely large scales and a deeply forked tail. It can reach more than eight feet and may weigh 250 pounds. It is a powerful migrator, related to the much smaller herring. It has a pouting appearance, due to its large protruding jaw. The high-metabolism silver king, hooked on a line, jumps high, pulls hard, and fights for a long time before agonizing at the mercy of the "big game" angler. As it is not edible, it is only caught for the pleasure of torturing, with endless refinements, one of the rarest and most beautiful creatures of the sea.

School of herring. *Herrings are important human food fish. They school in perfect unison, looking as if they had been trained for their journey.*

Slime for Speed

Most fish, like these herrings, have their scales—all along their bodies—covered with slime. Secreted through mucus glands in the fish's skin, the slime serves two functions. It acts as a lubricant to "grease" the fish's way through water with the least possible amount of friction. Being composed of long-chained molecules, slime helps stabilize water in laminar flow. Probably its most important function is that of sealing the fish and making it watertight. The skin of a fish is semipermeable, and without slime, the salt concentration in the fish would tend to equal that of the sea through osmosis.

121

Dolphins migrate across oceans at average speeds about double that of the cruising pace of the fastest cold-blooded fish.

Do Dolphins Defy the Laws of Hydrodynamics?

Performance of aquatic animals is usually estimated indirectly by calculating the drag forces on a fast swimming fish or a marine mammal and then figuring the muscle power necessary to overcome such drag. From computations of this type emerged Gray's Paradox, named for the famous scientist of animal locomotion. To explain the dolphin's ability to exceed speeds of 30 knots, this paradox pointed out either that the dolphin's muscles could produce ten times the power of terrestrial muscles, (man can develop about .024 horsepower per pound of muscle) or that their bodies generated com-

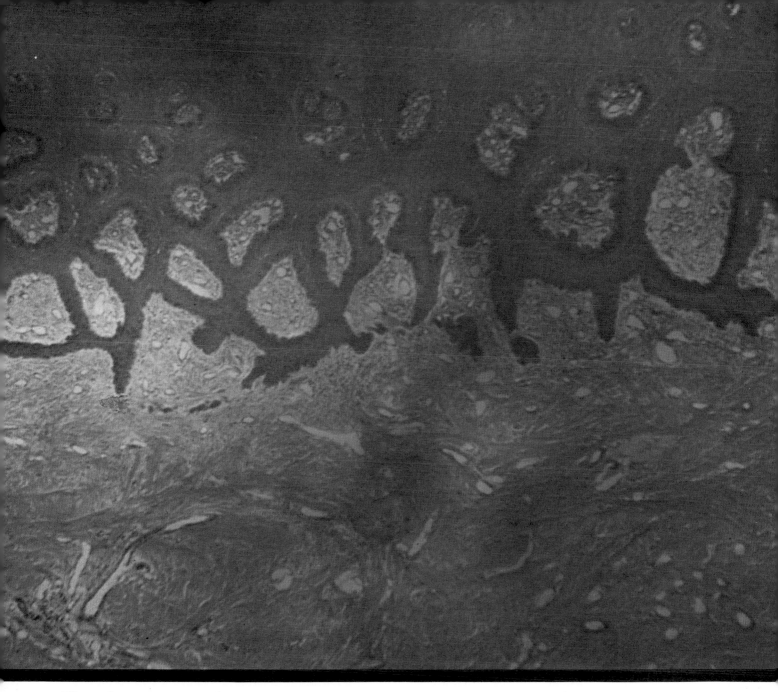

Three layers of dolphin's skin. Shown in this photomicograph at top is the epidermis, supported by layers below that can change the body contour.

plete laminar flow. Pure laminar flow has never been observed at such speeds and body sizes as those of the dolphins.

How could Gray's Paradox be solved? The dolphins could not be lying, so the drag calculations had to be inaccurate and grossly on the high side. Recent studies have proved that marine mammals have a flexible and pressure-sensitive skin that dampens boundary-layer turbulence. The dolphin's skin has three layers: a thin, flexible outer layer; a thicker middle layer containing channels filled with a viscous substance; and a stiff, thick inner layer. When the boundary layer tends to thicken in eddies, the skin is depressed and it is pushed out if that layer thins out. This passive system requires no energy from the dolphin and dramatically increases its performance.

Paddles

Sea lions and fur seals have retained four well-developed limbs from their terrestrial origin. They can still gallop on land faster than man can run. Underwater, they use their long, well-streamlined front limbs as powerful paddles while the rest of their flexible bodies and their hind limbs are mainly three-dimensional rudders. Antarctic penguins swim in a very similar fashion, using their wings as paddles. The sea turtle also paddles with its flattened front paws and leaves its hind paws trailing. Perhaps the most unique mammal propulsion system is that of the sea otter: its flat tail adds to the flattened hind legs to constitute a triple rear-propulsion device.

Fur seals. Above, these mammals swim gracefully, using their front limbs as wings and their hind limbs as rudders.

Sea otter. In the photograph at right, we can see that this mammal is well adapted to ocean life. Its tail is flat and has become a powerful propulsive aid.

124

Auxiliary Traveling Wave

Squids rival fish in their design for speed during brief runs, but they cannot compare with bonitos for long-range cruising. There is one field in which squids are unique: maneuverability. Capable of swinging their siphon nozzle almost 360°, they can aim the propulsive thrust of their jet and dart in any direction. But these members of the jet set have also adopted the "traveling wave" system as an auxiliary power transducer for low speed. On each side of their bodies is a horizontal muscular fin that can be controlled to undulate in both directions, so that they can shift in low or high gear as well backward as forward!

A / Moving forward. Here the squid's siphon is pointing toward the rear of the animal, and when water is forced through it, the squid will dart forward.

B / Moving backward. When the siphon is swiveled around toward the front of the squid and water is forced through it, the squid will shoot off toward the rear, its triangular fins acting as rudders.

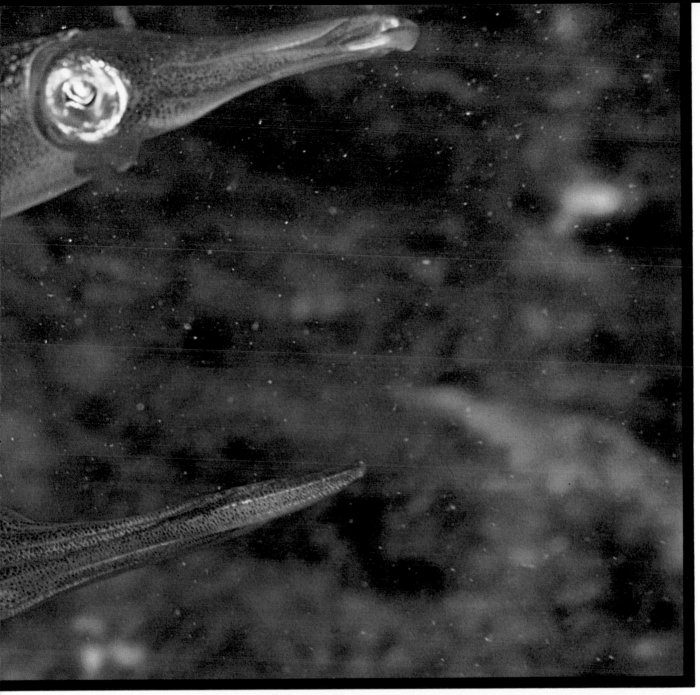

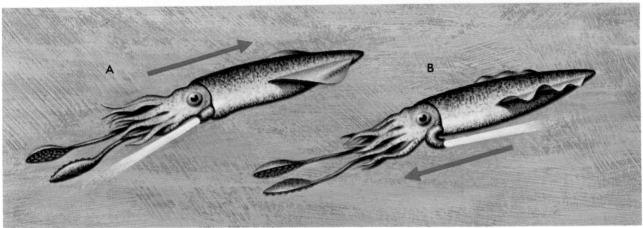

The Fastest Underwater Vehicle: Nuclear Submarine

With its periscope, snorkel, radar and radio antennae retracted into its sail and all hatches battened down, the nuclear-powered submarine *Sculpin* moves through the sea, its decks awash. Only the sail protruding upward from the hull's rounded, tapered cylinder breaks the boat's clean lines. The hull is a deliberate and careful imitation of the general body streamlining of swift cruising, large animals—mainly whales. Buoyancy of the submarine can be adjusted by filling or emptying ballast tanks, which more or less replace the mammal's lungs or the fish's swim bladder. Static and dynamic stability are achieved, as in most conventional submarines, by ballast in the keel, diving planes and rudders. The first great improvement was made at the power plant level, when the traditional, low-efficiency tandem "diesel on the surface" and "electric

undersea" was extended "diesel undersea" by the introduction of snorkels by the Germans in World War II. With the snorkel, the submarines were still breathing oxygen from the air, and remained in the sea mammal family. Then came the second, and maybe final revolution: the replacement of diesels and cumbersome electric batteries by a nuclear power plant. With unlimited range underwater, capable of producing oxygen from the sea, submarines became fish. Even with the mediocre efficiency of propellers as

Nuclear submarine. *Designed for fast, efficient movement through the sea, it is no coincidence that the submarine bears strong resemblance to some of the ocean's mightiest whales.*

thrust transducers, the new power plant is capable theoretically of pushing the submarines at incredible speeds, but the limitation comes from the difficulty of keeping longitudinal stability under control at very high velocities. Gyroscopic, airplane-type controls of the diving planes make fine vertical steering possible.

Chapter XI. Movement Shapes a Mode of Life

The form of an aquatic animal reflects its way of moving and its way of living. Thus, a slender, fusiform fish is likely to be a swift-swimming predator. Less streamlined creatures have adapted to a slower way of life. Similarly, a globular fish wallows, probably close to cover or otherwise protected from predators. The Portuguese man-of-war drifts on the surface with the wind. It has a sail, or float, which takes the wind the way a boat's sail does. Its trailing tentacles, catching what they can, slow it down and give it a better chance to sting and stun unwary creatures. The wind blows it anywhere just as it blows other drifters.

Crustaceans, crabs and lobsters, move in a cumbersome manner. Their armor doesn't permit flexibility of movement, so they lumber on all eight or ten legs, sometimes sideways, sometimes forward, their way of moving dictated quite strictly by their rigid coat. Consequently they seek food that is easy for them to get—slower-moving creatures or carrion. Or sometimes they dig. And burrowing animals display tools of their trade—limbs, heads, or appendages that aid in burrowing.

The skates and rays and other bottom-dwellers—flounders, goosefish, sculpins, and catfish—are adapted for a life on the sea floor. With their flattened forms, they can stir up bottom sediments and let the mud or sand settle on their backs, hiding them effectively. And, as we have seen, their shape enables them to "fly" virtually undetected along the bottom. Others have settled on the sea floor for other reasons. They may have no swim bladder, for example, so they sink to the bottom and have adapted fins that help them plod along.

Where a fish lives has a great effect on its form. Or, conversely, perhaps the fish seeks a place where its attributes are best suited. The trout and salmon are fusiform fish with all the characteristics necessary for life in a swift-flowing stream or in ocean currents. They have pectoral fins close behind their heads. They have pelvic fins on their sides halfway back along the length of the body.

The articulations of the vertebral column of most fish have a much greater degree of freedom laterally than vertically and it is exactly the contrary with marine mammals. Fish therefore, can swing in a horizontal plane and have developed a vertical caudal fin; they are poorly equipped to change rapidly the level at which they swim, and that is fortunate, because otherwise they could blow up their swim bladder. On the other hand, whales or porpoises can bend more in the vertical plane and have developed a horizontal tail; they constantly sound or beach, and this is a necessity for air breathing.

When a new creature is given a shape after a mutation, its propulsive capabilities and thus its way of life are a mere consequence of its anatomy.

The open-run shark is built for speed. With a fusiform body which cleaves the water with minimum disturbance this creature of the open sea can roam the world seeking smaller swift swimmers as food. In fact, because the open-ocean shark has no swim bladder, and its body is heavier than water, it would sink to the ocean floor if it didn't swim constantly and had no other equipment to counter the tendency to sink. But the large upper lobe of its tail fin provides some upward thrust; the pectorals, often long and broad, serve as planes that give it additional lift. The shark has yet another aid to float—an oversized liver. In some sharks the liver, containing lighter-than-water oil, reaches a weight of 1000 pounds. It is often 25 percent of the body weight and serves the animal as a buoyancy organ.

Awkward and Graceful

The manatee's awkward shape belies its grace in the water. Looking like a shapeless clumsy mass of flesh, the vegetarian manatee is a good swimmer. It has two short flattened forelimbs, just adequate to paddle slowly on the bottom or to help push back to the surface to breathe; but its enormous broad tail is incredibly powerful. It gives the manatee the ability to escape its very few predators.

Sea lions have streamlined, bullet-shaped bodies and an extremely flexible spine which enables them to make U-turns in less than a quarter of their length. Their front paws have become long wing-like flippers and they literally "fly" underwater, in the same free style that swallows use in the air. Like their relatives the seals, sea lions have a fur coat: such a hairy surface was thought originally to increase the friction drag and it probably does so, but to a very small extent. However it has recently been demonstrated that the fur padding acts as a boundary layer stabilizer, reducing turbulence in very much the same way as the triple-decked skin of the dolphin does.

With their speed and maneuverability, sea lions spend a minimum of time feeding. They enjoy freedom of movement, and often tease other animals or play with them.

Manatee. *This vegetarian mammal (above) eats the water hyacinth, a plant which, though harmless and beautiful, would otherwise flourish and clog the waterways around Florida.*

Sea lion. *This engaging mammal (at right) is the picture of grace, power, and joy in the water.*

133

Why Fish Face the Current

The ocean's waters are rarely still. They move in great currents that modify temperatures on a global scale. Fresh water flows from rivers that drain the continents. Eddies form on the fringes of major currents. Tides advance and recede as the moon circles the earth. Whatever causes water to flow, fish generally will face into the current. Mouths open and almost motionless, they only occasionally adjust a fin to maintain position. A fish facing a strong current can maintain itself in the same place with apparently less effort than would be required to swim at the same speed in still water. This can only be explained by comparing them with the large gliding sea birds, which are able to follow a ship for days without moving a feather. The fish take advantage of turbulence created by the current in water as the albatross benefits from eddies generated by the wind.

A fish facing into the current has a good

chance to capture a meal if it is moving directly toward the animal's mouth. Fish also extract oxygen from water that passes over their gills. Moving water, having been churned in air, is oxygen-rich. Of benefit too, in many cases, are temperature differentials that may enhance living conditions.

There is another advantage to facing the current: if a current were to hit the side of a fish, the animal would lose control and be carried along with the flow, perhaps into

Schoolmasters. These fish, which seem to be looking right at us, are a common species, growing to about 18 inches and weighing about three pounds. By facing upcurrent, as they are doing here, they don't have to work quite as hard for their food.

dangerous or intemperate locations. Facing away from the flow would similarly affect the fish's control of its speed and direction. Like more sessile animals, many fish let moving waters bring them the necessities of life, rather than pursue them.

Becoming Sessile

Barnacles start out life as free-swimming creatures. Only later, as they develop into a near-adult stage, do they settle down. Then they attach themselves to some solid substrate, like rocks, whales' bodies, ships' bottoms or wharf pilings. When they first hatch out of eggs, they resemble their crustacean relatives, the crabs, lobsters and copepods.

At this nauplius stage they swim about freely. After several moults they change to the cypris stage. In this stage, the barnacle is in a bivalve shell with a muscle to hold the two valves or shells together. It also has developed a pair of compound eyes as well as a cement gland which it will use to glue itself to something solid. As an adult, the barnacle uses that gland to stick itself onto a solid substrate. It depends on the movement of the water around it or the movement through the water of the substrate it is attached to for food. Its bivalved covering has added plates that seal it in except at one end. From that end the legs protrude, kicking plankton into their mouths.

Some barnacles are minute in size. Others, such as those found on whales are rather large, reaching a size of six inches. A few species are parasitic on crabs. Some others are commensal with whales, turtles or a few fish.

Starts life swimming. Above, larval stages of barnacles are many. In this stage the larval barnacle is still free-swimming and is about to change its way of life.

Sessile colonies. In its adult and more familiar form, the goose barnacle shown at right is found in colonies, securely attached, unmoving and almost immovable.

Angelfish

The tall narrow form of the queen angelfish governs its life style. Fish of this family and others that have laterally compressed bodies maneuver easily. Their flat shapes enable these fish to slip into tall and narrow crevices in coral reefs to find protection from larger predators. Here they can also find the tiny morsels of food their small mouths can take in. Their flat bodies can fit into horizontal narrow crevices that they sometimes penetrate after leaning completely on their side to swim parallel to the bottom.

"These fish and others that have laterally compressed bodies are extremely maneuverable. They slip into tall and narrow crevices in coral reefs, finding protection from predators."

Flatfish

The flatfishes like this sole and its relatives, the flounders and halibuts, are in fact laterally compressed fish although at a glance they give the impression of being dorsoventrally compressed. The swimming movement of a flatfish is much the same as that of other fish. It propels itself by bending its body from side to side in a traveling wave, moving horizontally rather than vertically. In its larval form it is no different from normal fish. It lives at the surface of the sea with one eye on each side of its head. Then one of its eyes begins to migrate to the other side and the shape of the fish gradually flattens. Finally it sinks to the bottom. Its eye having migrated completely, it lies over on one side and spends the rest of its life that way. Flatfish have a mode of life which is almost completely governed by their strange shape and position. The advantage they have is being able to hide on the sea bottom, covering themselves with sand.

139

Sea Anemones

Sea anemones, relatives of the more mobile jellyfish, have become shaped to fit the sedentary life they live. Their form is stalk-like, resembling a plant; in fact they are often mistaken for plants. They can move if they must, but usually they remain in one place for extended periods of time. Generally they live a quiet and sessile existence, taking food from water that passes by them. When they do move, it is with great awkwardness. To move some must release the hold their pedal disk has on the substrate; they then fall over on one side, twist around, and re-attach their pedal disk in a new location. Others merely inch along by snaillike movements of their pedal disks. Despite this inability to move along freely and easily, the sea anemone is very well protected. Its stinging tentacles, which stun its prey, can also be retracted when the animal is not feeding. Delicate, decorative looks are more than compensated for.

Little Fish on Reef

Why do so many little fish live around reefs? It may be because it is an ideal place for small fish. On a reef, whether it is made up of coral or of geological rock, there are many tiny cracks, slots, and crevices, which serve as cover for small fish but not for large ones. Vast numbers of microscopic plants and animals around these reefs serve as food for the small reef fishes. Consequently their mode of life does not require that they be great swimmers; to avoid capture they must only be able to dart into a nearby hole. For instance, in this tropical reef, the clownfish remains close to the anemone's stinging tentacles. In this marvelous relationship the clownfish may lure prey to the anemone, and the anemone protects the clownfish from predators. Dazzling color and variety abound in and around coral reefs, and many of the creatures living in these communities are camouflaged and assisted by their surroundings.

Chapter XII. The Art of Motion

I plunged into the sea and I forgot my load
I reached for a freedom that stunned and froze my heart

        Freedom in three dimensions, cold and thick
        And liquid embraces that stalled my skin

A dash of fluttering broth
A pulsing crystal froth

        Spirit of the sixth sense
        Speed spreading insolence

The cavalcade of shrimps
Flesh belt's extravagances
Rainbows of darting squids
Anonymous dances

        And vertical gold leaves
        Heading into the stream

Instinctive snouts gaping for oxygen and prey
Throngs of rhythmical fins beating mindless pathways

        Then dolphins, seals, and whales stole fire
           from the sun
        To playfully invade the automatic sea
        In their burning entrails our
           powerful cousins
        Presented their mother with songs, luxury,
           and fun

But since I discovered the key to fluid dreams
The warmth of the whale's blood simmers deep in my heart.

        While diving birds regret their weight
        And leaping fish plane in the wind
           Moving elsewhere
              Elsewhere for hope.

PART TWO

The Ocean World of Jacques Cousteau

Part II:
Attack and Defense

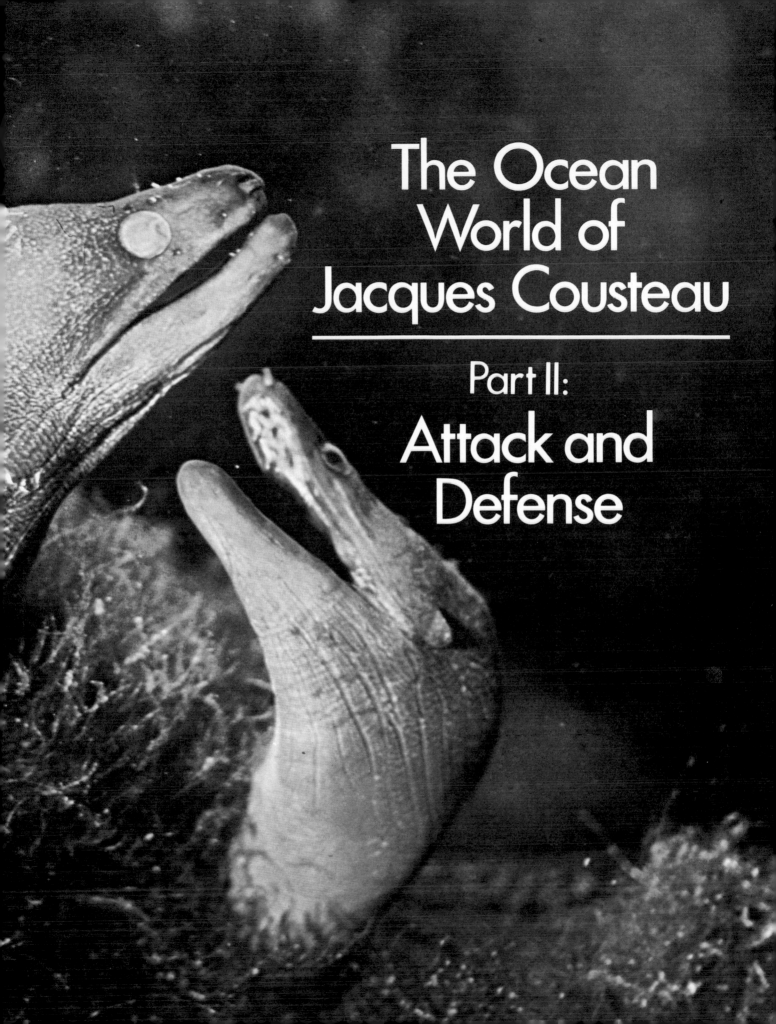

The Ocean World of Jacques Cousteau

Part II:
Attack and Defense

The **sheepshead,** aggressively displaying his strong teeth, does not use these as weapons of attack against other fish. Instead, it uses them to penetrate the hard shells of its sedentary prey—shelled snails, crabs, bivalves, and sea urchins.

Contents: Part II

In the sea, as on land, (wherever man has not interfered with natural processes) attack and defense keeps the life pyramid in balance. IT'S A HARD LIFE but all is not blood and gore in the "struggle for existence." The equilibrium between offense and defense, provide opportunities for the pleasanter sides of life: mating, childbearing, playing.

But many of the sea's creatures still possess BAD REPUTATIONS (Chapter I): the "killer whale" and shark because of their power and speed and carnivorous appetites, jellyfish because of their occasional stings, rays and eels because of their bizarre appearances, billfish because of their ability to fight to the finish, and other animals like the sea snakes and scorpionfish because of their potentially dangerous venoms.

Although coldblooded marine creatures need less food than the warmblooded, catching a prey is very tricky in a three-dimensional world. KILLING FOR A CAUSE—HUNGER (Chapter II), is the first order of urgency in the sea as on land. In some cases, like the feeding frenzies of sharks, this killing can take the form of terrifying orgies, lasting for long periods of time with countless smaller fish consumed. But the march of the poisonous cone snail upon its hapless prey, the feathery lure of the anemones, the lovely dive-bombing of a seagull, while less melodramatic, are just as effective.

Animals employ camouflage for both offensive and defensive purposes. COLOR ME INVISIBLE (Chapter III) is one of the most successful of military tactics. What it involves can be likened to playing with light: countershading, disruptive patterns, color changings which can occur instantly or over a period of time, false leads (like the butterfly fish's false eye near his tail), mimicry (which in the case of the Sargassum fish involves growing various protuberances which help them more closely resemble a floating weed), and distracting stripes, which can be either vertical or horizontal, but always confuse the predator.

When going into combat primitive man probably armored himself with a few skins wrapped around his arms. Gradually evolved were bucklers and greaves and helmets, until by the dawn of the Renaissance the European man-of-war was entirely encased in metal from head to toe. He then shucked himself out of these shells, preferring that his engines of war largely

carry the armor while his personal movements retained some flexibility. Each one of these stages is reflected in the sea in one or more of the many animals who are LIVING IN ARMOR (Chapter IV). Some have armor that grows with them, others must change armor as they grow, and still others live in a type of borrowed armor which they, too, must change as they increase in size.

"For he who fights and runs away/May live to fight another day." Here is the idea behind STRATEGIC WITHDRAWALS (Chapter V). By jetting or gyrating out of the reach of a more powerful foe like scallops or nudibranchs, by hiding behind rocks or behind the surface (by breaking into the air flying fish in effect "hide" behind the sea's surface), by closing up shells or digging into the ground, like certain molluscs, and a hundred of other tricks, the creatures of the sea attempt to win another chance. Some even choose to live in exquisite castle-like limestone structures.

If battle cannot be avoided by escaping or hiding, an animal has no choice but to employ OFFENSIVE DEFENSES (Chapter VI). Almost all man's own weaponry has been inspired by nature: smoke screens, tear gas, guns, hammers, poisons, etc. Like Greek warriors insulting the enemy before fight, some animals even use sound; the great roar of an adult sea-lion puts a younger rival in an uneasy frame of mind before a fight even starts. Others are less vocal and less aggressive— they simply taste terrible.

When battles do occur between members of the same species, it is almost always FIGHTING FOR TERRITORY AND SEX (Chapter VII). Given these incentives, some of the smallest and least combative creatures turn as fierce as titans of the sea like the elephant seal. Most of the time, however, the throwing down of the gauntlet is normally enough to allay any physical combat.

The octopus, the moray eel, and the crayfish are a case of ANCIENT ANIMOSITIES (Chapter VIII). Probably because ages ago this group competed for the same territory or food, such a trio finds itself locked by evolution into a relationship of elemental feud which lasts as long as the three species themselves endure. Other examples of Hatfield and McCoyism in the Animal World: the dolphin and the shark, the penguin and the skua, and the beluga and the "killer whale."

But all this attacking and defending ebbs at certain periods in the sea, and there is A TIME FOR PEACE (Chapter IX).

Introduction: It's a Hard Life

"Nature red in tooth and claw"?

Civilization has the fundamental ambition of introducing a degree of order into the primeval "struggle for existence." But to this day it has only succeeded in drawing a curtain between its harsher manifestations and certain of our human sensibilities. Most of us have never visited a slaughterhouse or a fish cannery or a 24-hour-a-day illuminated chicken-and-egg farm. When we fall in love we rarely have to dispose of our rivals by means of a billyclub. When a neighbor turns nasty and tosses beercans into our backyard we call the police. When the police collude with the neighbor we call the state or Federal authorities. When we play rough games we institute rules and time-periods and an umpire.

But the classic, bloodstained struggle simmers right along in easy view if we care to look for it. We need protein just as urgently as the shark needs his mouthful of red snapper even though we are fed by the farms and stockyards and trawlers of this world. The newspaper reminds us daily that lust and territorial dispute and injured vanity and greed turn into acts of savagery and murder in the best-regulated communities. At a more general level, nations hurl themselves into vast conflagrations which consume millions of men and irreplaceable treasure for causes which—in the past several hundred years at least—bear no sensible proportion to costs. Man has lots of shark in himself still, as almost all of us find out at least once or twice in our lifetime.

The creatures of the sea do not write poetry or paint pictures. Their lives are more obviously determined than ours by the basic quests of life: for survival, for food, for a mate, for a territory—for play. And in pursuit of these quests they have developed over the evolutionary eons offensive and defensive weapons in nearly every conceivable direction. Man can find a precursor of almost all his primitive or refined armament in the sea. There are animals that use the analogues of swords, spears, bows and arrows, nets, electric cattleprods, camouflage, armorplate, speed, poison—often two or more instrumentalities in dazzling combination. Moreover, advanced animals utilize tactics and strategies. (A lioness will stampede a herd of zebra toward her invisibly crouching mate; the stratagem is roughly that employed by Napoleon at Jena.) Barracuda "herd" schools of smaller fish. Dolphins hunt in packs. A male "killer whale" will lure a ship away from his vulnerable family. In a properly conducted seal-rookery (one not too overcrowded) there is a strictly enforced order in the access of males to females. No matter how exigent his passion, the young male must bide his time or get badly chewed up. It is all the struggle for life, for survival. With the notable exception of the explosives man concocts in his laboratories (culminating in the biggest bang of them all—the atom bomb) man can learn everything he needs to know about the principles of attack and defense from some creature in the sea.

Still, in the sea, as elsewhere in nature where it has not been contaminated by civilized man's ecological irresponsibility, there are balances struck between attack and defense. For example, predators survive only if prey also survive. If too many sea otters devour too many urchins and abalone, the otter population soon suffers from the effects of starvation. (Then, of course, the urchin-abalone population recovers, in turn triggering a resurgence in otters. The mathematics of such particular prey/predator relationships have been worked out in research studies.) Theoretically a too-successful predator will ensure his own extinction. The thought that this might be man's destiny is intolerable, but we will need a great amount of vigilance, imagination, and sacrifice to avoid such a fate. In general the world we see and live in comprises a tapestry of delicately interlocked strands like the otter/abalone pair, rising and falling in innumerable consonant rhythms. The efficiencies of attack and defense systems have become attuned to each other over the 2 or 3 billion years that life has evolved on our planet.

It is this automatic "tuning" of the natural world, these dynamic balances between prey and predator, aggression and withdrawal, that Western civilization is disrupting—with the consequence that the job of restoring an order to nature is now man's alone. Man is an animal, and the "nature red in tooth and claw" side of his animal nature is never far from the surface. Yet with his brain, and his languages, and his prehensile hands, man has liberated himself from most of the laws which rigorously limit the possibilities available to the rest of the Animal World. In the field of weaponry he has borrowed a tiny chunk of the sun's own fire for his thermonuclear devices; armament cannot go much farther than that. If he wants to, he can reduce earth to a nightmarish desert, or blow it up altogether. But why should he do these things? How much more in harmony with the other side of his animal being—his instincts for survival, for mating, childbearing, playing—that he turn the marvelous tools of intelligence and analysis which produced the ultimate weapon of attack to the fabrication of the ultimate defense: Peace—human societies living at peace with one another, a human species living at peace with its world and with the other inhabitants of its world.

Jacques-Yves Cousteau

151

Chapter I. Bad Reputations

To restate an old saw—when a man bites a fish, that's good, but when a fish bites a man, that's bad. This is one way of saying it's all right if man kills an animal, but if an animal attacks man, the act is reprehensible. The animal is labeled "killer," something to be feared, hated, shunned, punished, even killed by man.

Stories of vicious attacks, true or not, give some sea creatures a bad reputation, deserved or not. When the tales are true, the animal is feared or disliked because it hurts man—it bites, stings, stabs, or infects him.

> "There are a few animals that have won themselves a bad reputation even though they have little or no effect on man. Their rating comes from man's interpretation of their attitude toward lower animals —for example, feeding in a so-called savage manner."

But in most cases a sea creature attacks only to defend itself against man's predation. Rarely does it threaten man without cause. Untrue stories about the "ferocious" behavior of a sea creature often arise when man knows little about the accused. The animal's mystery gives rise to wild tales that condemn it to man's list of "bad" creatures. But how dangerous are those sea animals with bad reputations? A few actually kill. A few maim. Some are poisonous when eaten by man. Most sting, stab, or poison and cause mild to severe discomfort to man. The poisons of sea creatures may affect either man's circulatory system or his nervous system. Those affecting the nervous system are usually the most dangerous because they may

shut down the brain's signals that keep bodily functions like breathing continuing. But man is one of the larger beings that sea creatures encounter, and these poisons usually can't kill him. Very often these poisons are used defensively against predators and offensively in food gathering.

There are a few animals that have won themselves a bad reputation even though they have little or no effect on man. They have won their rating through man's interpretation of their attitude toward lower animals. These animals have been seen feeding in what appears to be a savage manner. But this behavior may perhaps be comparable to a man tearing the flesh off a chicken leg with his teeth. Some animals of the sea often seem to be feared by other sea creatures. A large predator that swims in leisurely fashion until it spots a school of small fish may provoke or seem to provoke fear in the school, and the school may suddenly and dramatically maneuver to escape. It then seems to us that the predator has a bad reputation among its fellow inhabitants of the ocean.

Killer whale. *English-speaking people have dubbed the orca, largest of dolphins, the "killer whale," and until very recently it has led the list of sea monsters. Its fearsome reputation stems from the fact that their stomachs have been found to contain remains of walrus, seals, birds, and dolphin. Whalers have also reported that orcas drive group attacks on large whales, biting off lips and tongue; but this probably happened only to helpless harpooned whales. In fact, orcas eat almost anything, mainly fish and squid. The orca is a sleek, agile, and intelligent animal, averaging 20 to 30 feet in length. They travel in groups of 10 or 12, always led by a large male. In spite of its reputed bloodthirstiness, the orca is a rare animal, not nearly so numerous as that very successful predator the shark. There is little doubt that the orca's powerful jaws and sharp interlocking teeth could tear a man apart, but there is no record of such an attack. There is still much to learn about orcas, but it is obvious we have misjudged them by labeling them deliberate killers.*

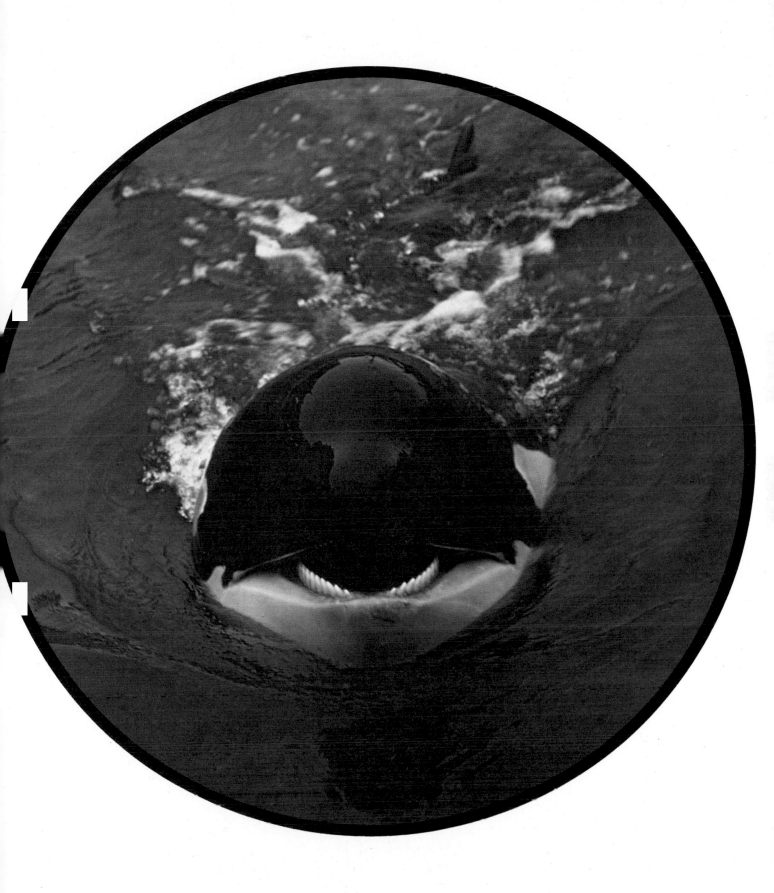

Reef shark. One of the requiem sharks which includes some species known to possibly attack man, this shark is common in coral reef areas of the Caribbean. Obviously, any inquisitive behavior of a shark is bound to be felt as a threat by divers. As a consequence, but without reason, reef sharks are feared.

Toothy Threats

The word "shark" strikes fear into the hearts of many men. In some cases this fear is justified. In most it is not. Of the 250 species of sharks currently recognized, only about 35 to 40 may eventually be dangerous to humans. But as Dr. Henry B. Bigelow of Harvard has said: It's better to get out of the water if you're uncertain. The more than 200 species of sharks not considered a peril to humans are ill-equipped, physically or

The sand tiger. With its row on row of sharp, curved teeth, the sand tiger is one of those sharks that is supposed to menace man, though not in North American waters. Sand tigers generally eat fish, crustaceans and squid. Sand tigers are one of the few predators that attack the bluefish.

temperamentally, for such activity. Some have flat-topped teeth, others are too sluggish, a few are too small, and many just aren't interested in man as food or as a threat. So when they do encounter man, they go the other way—and so does man.

"Most sharks are no peril to man. When they do encounter him they go the other way— and usually so does man."

Stingers

Jellyfish have a bad reputation, which is deserved by some species but not by others. Pelagia, have tentacles that carry poisonous stinging cells named nematocysts. So do many other jellyfish. A few, like the white moon jellies that range the oceans of the world, are harmless. Some, like the red-tinted Cyanea, found in temperate waters, and the sea nettle of tropical waters and those of the eastern United States, are usually passive but will sting and release their venom if intruded upon by man or sea creatures. In man their sting causes at least burning and itching of the skin and at worst swelling, redness, and breathing difficulty.

The Portuguese man-of-war secretes poison powerful enough to do serious harm to men.

Under a pink blue gas-filled float hang many filaments reaching down 40 to 60 feet and including nematocysts, or stinging organs, capable of injecting poison. The man-of-war stings indiscriminately to feed and thus to survive. Most predators dare not touch it and only a few seabirds are known to eat it. In laboratory experiments, conducted at the turn of the century, Professor Portier, working with Prince Albert I of Monaco, injected repeatedly decreasing doses of the man-of-war venom to a dog that died. Examination disclosed that the dog was allergic to the venom.

Pelagia. These jellyfish, above, have nematocysts with which they sting their prey.

Portuguese man of-war. At right is the Portuguese man of war, who uses its stinging poison to stun its prey.

Manta Ray

These powerful animals have an enormous wingspan—but they are totally harmless. Still, mantas are called "devil rays" by many fishermen, and Cuban fishermen have superstitions about them (especially about their hypnotic powers, their habit of jumping out of the water onto fishing boats, and other threatening actions). However, fishermen who harpoon one of these giants soon discover its strength. A manta can demolish an ordinary fishing boat in a matter of minutes, but this is a normal fight for survival. The largest observed specimen of this animal had a wingspan of 22 feet and weighed almost two tons. The two fleshy protuberances on either side of the manta's head are believed to funnel water into its mouth, and with the water, the small fish and plankton it lives on. Far from dragging sailors to watery graves, rays are content being left alone, occasionally jumping clear of the water three times in a row.

Stingray

It is not difficult to understand why the stingray has developed its bad reputation. It has a fearsome, whiplike tail longer than its body, and near the base of this tail are one, two or three flattened spines with small, sharp teeth—coated with venomous slime which can bring serious injury or even death to man. But our misconceptions center on the manner in which the stingray uses this formidable weapon.

A stingray leads a quiet life on the ocean's floor and will never attack a man. If approached, it will flee. The stinger is used only as a defensive weapon, not as an offensive one. Its position on the tail enables the ray to sting an enemy above it. When threatened, the stingray will whip its tail around until it finds its attacker. The stinger is never even used to obtain food for the ray; the ray feeds by sucking molluscs and crustaceans into its mouth. If a diver or swimmer steps on a stingray's poisonous spine, who is to blame?

Barracuda

The barracuda's razor sharp teeth and powerful jaws coupled with its ability to strike its prey with lightning speed have given it its reputation as a killer. Its mechanism for killing, to be sure, is extremely well developed. When it has to feed, for instance, it slashes through a school of fish with its teeth glinting, ripping and tearing, leaving dozens of dead or dying fish in its path. It then returns to feast. It is an inquisitive animal, and its habit of hanging almost motionless in midwater watching swimmers and following their every move has furthered this reputation. The fear the barracuda strikes in swimmers, divers and fishermen, however, is not justified.

Actually this sleek powerful fish, which may grow to a length of six feet or more (some have grown close to 12 feet), has never been recorded attacking or hunting a diver except by accident, when it is landed in a diver's or fisherman's boat after it has been caught. When a diver, even unarmed, swims toward a barracuda, the latter dashes away—but not very far indeed—and soon it is back behind the diver, close to his feet fins. As the impression of being followed by such an enigmatic predator is somewhat disagreeable, the wary diver turns around and threatens the fish away . . . for a few seconds. This game of intimidation can last for hours with no apparent lassitude in the barracuda.

Barracuda. It is easy to see from the photograph above, why the barracuda has its reputation as a killer. Its sharp teeth and streamlined shape make it a formidable foe.

Marlin. This fierce-looking fish (opposite) may thrash about for long periods of time at the end of a fishing line. When desperate, it will attack anything.

Billfish

Billfish, like this small, black-striped marlin, are extremely dangerous when hooked and played on fishing tackle. This marlin has been played almost to exhaustion, and although the hook is holding fast in its mouth and the fishing line is being held taut to prevent the fish's escape, it still hasn't been landed. When nearly exhausted, marlin and related swordfish may turn and attack men, boats, or anything else near them. They have stabbed men even after being landed and while thrashing about on the deck of a fishing craft. In an underwater encounter a few years ago, a swordfish skewered the submarine *Alvin*. The fish died, *Alvin* recovered.

> "In an underwater encounter a few years ago a swordfish skewered the research submarine *Alvin*. The fish died, *Alvin* recovered."

Arms and Armless

Some of the most venomous creatures in the sea are the sea snakes. When they bite their venom paralyzes the nervous system and the victim soon dies of suffocation. But the sea snake, including an olive brown variety found in the Great Barrier Reef off Australia, is always unaggressive. It is often said by natives that sea snakes have a small mouth and could only bite man's tender skin at the base of the thumb. This is not true. These snakes can bite anywhere, but they only do so if seriously disturbed. In the Persian Gulf many pearl-divers, who did not wear goggles, have been killed by sea snakes. Sea snakes range all over the tropical Pacific and Indian Oceans. With the completion of a projected sea level canal through Panama, however, Pacific sea snakes may find their way into the Caribbean and Florida.

Legend tells of octopods wrapping their eight arms around men and squeezing them lifeless. In fact, however, these creatures are shy and retiring and quickly back away from an approaching diver or swimmer. But if they are attacked by their ancient enemy, the moray eel, or by man, they can and will use their strong tentacles to resist capture. And if sufficiently provoked they will bite with their powerful beaks. The blue-ringed octopus is a unique type of octopus. It is rarely larger than four inches long, but its bite is often fatal. Beachcombers of Australia are therefore warned that these "cute" animals are deadly.

Diver and octopus. The octopus (left) is not the man-crushing beast it is reputed to be, and in fact, would rather back away than fight.

Sea snake. Shy and retiring, the sea snake (above) is nonetheless one of the sea's most deadly creatures.

163

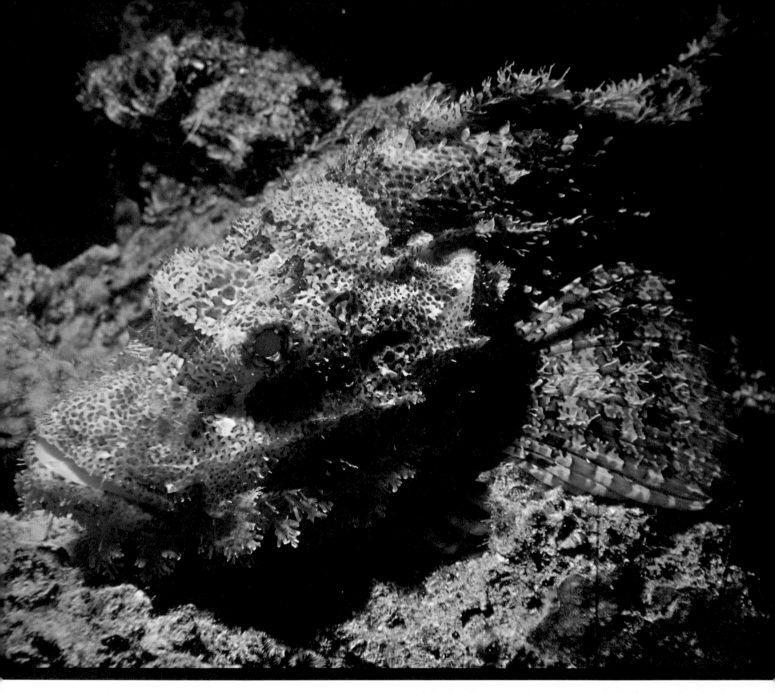

Scorpionfish

Scorpionfish are most deserving of their bad reputation. Some of their 13 dorsal spines carry the deadliest poison of any fish—poison lethal enough to kill a swimmer or beachcomber in two hours. These spines indiscriminately sting anything that touches them. What is worse, scorpionfish so closely resemble stones that they are almost unnoticeable, and it is all too easy for a person walking in shallow water to fall victim to one of them.

Their disguise is a part of their method of attack; when a smaller fish unknowingly approaches them, they quickly swallow it.

"Most deserving of their reputation, these fish indiscriminately sting anything that touches them. What is worse, they so closely resemble stones that they are almost unnoticeable on the bottom."

Crocodile

Looking like a fabulous creature from the Age of Reptiles, the crocodile has a fearsome reputation. These great reptiles, some of which reach more than 20 feet in length, live in tropical and subtropical fresh and salt waters. Large species, like northern Australian crocodiles, may grasp a large mammal like a cow and spin over and over, submerging the victim until it drowns. They can also stun a victim with their powerful tail. Their large teeth are one of the most powerful weapons in existence. Alligators, their relatives, found only in the United States and China, are less aggressive. The saltwater crocodiles of southern Asia and northern Australia as well as the Nile crocodiles are supposed to be especially vicious. In fact, crocodiles spend their lives hidden in very turbid water. On land, or in clear water, they are fearful and cautious. If man has ever been their victim, it was by accident, and long ago.

In addition to being able to attack other fish (for whatever reason) with remarkable effectiveness, the bluefish, once it has been hooked by an angler, fights furiously and jumps violently at the end of a line. When finally caught and in the boat, it has been known to grab for and bite off a fisherman's finger. Conversely, the bluefish can be made into a good meal itself once it has been caught, scaled, gutted, and broiled.

Bluefish

Bluefish are known as one of the sea's most bloodthirsty fish. These fast-moving fish do not just kill to eat—they often kill for no apparent reason. Long after their hunger is gone, they continue to slaughter, leaving shredded, half-eaten fish in their path.

School of bluefish. (Upper left.) As they travel near shore on their annual movements up and down the east coast of the United States, bluefish make the water boil with their activity. It is one of the few marine creatures that eats when it isn't hungry.

Evidence of the bluefish's destructive powers. Below and on the opposite page, these yellowtail flounders were brought to the surface shortly after a school of bluefish had traveled through, and as we can see, nips have been taken out of them. Many like them were found in the haul, and surely the ravaging was not only for the purpose of obtaining food.

Moray Eel

Moray eel. *This open-mouthed, fierce looking reef fish would rather turn away than fight. Once they do feel threatened enough to fight, however, they can do great harm.*

Moray eels have long had a reputation for being attackers. It seems to have begun in Roman times—historians tell of Nero grotesquely throwing slaves into water-filled pits of moray eels just to amuse some bored, aristocratic friends whose pleasure came from seeing humans being eaten alive. In point of fact, moray eels are retiring and would rather hide than fight, and so if this tale is true, Nero must have either given them no place to run to when the slaves were cast in, or starved them until they were desperate.

These four-or-five feet long (sometimes reaching 10 feet) eellike fish will only attack when threatened. Morays threaten but won't bite unless provoked. Once they bite, however, they can do considerable injury with their strong jaws and many sharp teeth. Their manner of breathing requires them to open their mouth rhythmically every few seconds to pump water over their gills, and this action is often mistaken by divers to be a threat. Attacks on men, however, are usually the result of divers' probing into the holes in rocks and coral that shelter morays. The bite of their nasty teeth is luckily not poisonous, but wounds must be quickly treated to avoid infection. Fishermen catching a moray are advised to crush the animal's head at once.

Their snakish, muscular bodies are perfectly suited for living in and around tropical reefs where there are nooks and crannies in which they can search for their meals and places to hide, which they do during most of the day. One of the reasons the moray and the octopus are such great enemies, in fact, is that they both seem to seek out the same type of resting places and all too frequently run into one another.

North American Lobster

It is not difficult to understand why man has developed some misconceptions about lobsters. The lobster's claws are strong, and no one who has been pinched by a lobster is likely to forget it. The big, heavy claw with blunt teeth is called the crusher claw. This is used by the lobster to crush its prey and to fight with other lobsters. The other claw, with sharp, teethlike projections, is the rip-

Crusher and ripper. The lobster's claws, called the crusher and the ripper, can be formidable weapons, but the lobster will retreat if it is approached by a diver.

per claw and is used to tear apart food that has been crushed by the crusher claw. It is interesting to note that if a lobster is approached by a diver, it will retreat. Only if it is cornered in its hole will the lobster use its powerful claws to defend itself.

Chapter II. Killing For a Cause—Hunger

Man sometimes views the sea as a tranquil world, shielded from the violence of the land. Not so. Death is a way of life in the sea, also, and nearly all activity of coldblooded marine

> "Death is a way of life in the sea. Nearly all marine-animal activity, both offensive and defensive, is motivated by the attempt to allay hunger or avoid predators. In the sea as on land, an appropriate motto for life is 'Eat and be eaten'."

creatures, both offensive and defensive, is motivated by the attempt to allay their hunger or to avoid falling prey to others in the sea as on land. An appropriate motto for life could well be "Eat and be eaten." However, the quest for food is slightly less drastic than on land, partly because fish need less food than birds, for example, and partly because sea mammals, who have great appetites, are so superior to the others that they spend very little time filling their stomachs.

> "Although violence and killing occur continuously in the sea, almost no animal kills unless it is feeding or fighting for survival. Prey and predator live most of the time in harmony. But whenever the predator gets hungry, the prey is gobbled up."

Hunger is one of the most powerful animal drives. To satisfy this drive, a great variety of eating habits have developed.

Tiny crustaceans and the larvae of larger animals eat microscopic plants and animals.

In turn, these diminutive hunters are preyed upon by larger hunters. And so it goes on, on up the food chain: larger animals eat smaller ones, only to be killed and consumed in their turn. And the largest creatures if unchallenged during their lives, are part of the food cycle too—they fall victim to tiny scavengers when they die.

But although violence and killing occur continuously in the sea, almost no animal kills unless it is feeding or fighting for survival. And animals usually do not kill just because food happens to be available. In fact, a fish may live safely most of the time near predators. But when the predator gets hungry and feeding conditions are right, it may be gobbled up. Hunger then is the first requisite for feeding.

The physical conditions of the oceans influence the desire to feed. These include the temperature, salinity, and acidity of the water, which vary with season and from place to place. Light plays a vital role. Deep creatures come to the surface at night, feed upon the shallow liquid meadows, and descend back to the depths at dawn, chased by daylight that they cannot endure. In the reefs, some animals hunt only under cover of darkness. But the great majority eats at dawn and dusk, during brief "feeding furors" in which birds, fish, squid, and mammals participate.

Patrolling shark. Patrolling the margins of coral reefs, sharks take advantage of the food supply found there. These magnificent creatures are usually patient hunters, waiting until an easy meal presents itself. They often feed on injured, dying, or dead animals, which have escaped other predators. These sharks show no preference for feeding either by day or by night. They often gorge themselves and then may not eat again for days.

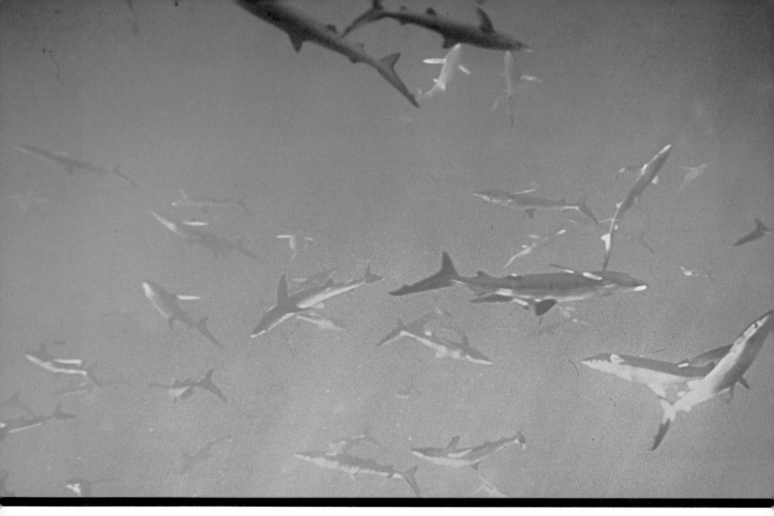

Feeding Frenzy

Blue sharks. *Attracted by the scent of food, these great fish swarm in the open ocean.*

Sharks are unpredictable. Sometimes they may swim casually about a diver for hours without showing any interest in him, and on other occasions they may behave erratically, the ambient field becoming electric, as soon as a diver enters the water. Sometimes sharks flee from an unarmed, unprotected swimmer, and at other times they may deliberately crash into the steel bars of anti-shark cages and bite furiously at them in unprovoked attack. The blue is usually a solitary hunter, but when the scent of blood is in the sea, many will appear as if from nowhere, like vague shadows suddenly come to life. They circle cautiously sometimes for hours until they are sure that there is no danger. Then one of the circling pack rushes the intended prey, brushing or bumping it. If the object seems edible and harmless, the boldest of the cautious group approaches for the first bite. Then a feeding frenzy may begin. Other times there may be blood in the water or a struggling wounded fish or even a shark attacking and no frenzy will occur. What initiates the shark's erratic behavior is not precisely understood. A number of factors are probably responsible for this behavior and each stimulus by itself may not be enough to initiate the frenzy. One thing, however, is sure—once a feeding frenzy begins the behavior of those active sharks stimulates all the sharks in the vicinity to become many times more aggressive than normal. The one, odd conclusion we can come to is: the better acquainted we become with sharks, the less we know them— one can never tell what a shark is going to do.

172

The Shark's Eversible Jaw

Sharks *can* bite a very large object, such as the side of a whale, with their dorsal sides up. They do so by dropping their pointed snout and their lower jaw, and by snubbing up their upper jaw in a grotesque manner. This snubbing action opens their mouths so wide that their jaws are nearly vertical, and the huge cavity of the fish's mouth is revealed, as are its rapierlike teeth. In addition, strong muscles allow the upper jaw to be thrust outward to grasp the flesh and then rotated downward in a cutting action. This efficient feeding system enables the shark to partake of a wide variety of foods.

When it is opened wide, the shark's mouth looks like a huge steel trap ready to be sprung. When a shark takes a bite of a large animal, like a whale or dolphin, it clamps onto the animal with the great jaws and sinks its teeth into the flesh. Then it seems to go into convulsions, violently wriggling its body from head to tail. Its razor-sharp serrated teeth are twisted from side to side, and they scoop easily through the captive's flesh. This awesome spectacle is over in an instant. The bite of a shark leaves a cavity in its prey's body.

The mouths of most fish are at the extreme forward end of the body, but that of the shark is not. The shark's mouth is placed well on the underside of its head, so much so it would seem sharks would have difficulty eating anything but prey smaller than the aperture of the mouth. In fact, Aristotle thought that sharks could not bite into their food from their normal swimming position, and he postulated that they roll over to eat.

The movable jaws of some sharks enable them to bite chunks of flesh off a victim too large for one bite.

Always On the Prowl

On an expedition to the Red Sea to study and tag sharks, *Calypso* divers entered the water, protected from the unpredictable animals by antishark cages, such as the one shown above. To bring the sharks close enough for tagging, the water was churned with small bits of fish, whose scent attracted the lurking sharks. Soon sharks were all around, circling the cages. The atmosphere was tense; although the scent of food was present, there was nothing for the sharks to eat. Suddenly a diver noticed a large red snapper cruising by and decided to spear it for additional bait. But before the snapper could be retrieved by the diver, it escaped from the spear. Suddenly the spell the sharks seemed to be under was broken. A shark broke away from the pack and rushed the injured snapper, chasing it into one of the antishark cages. Then the sharks abandoned their usual caution and hurtled into the steel bars of the cage trying to get at the wounded fish. The snapper escaped from the cage but not from the sharks. In an instant they were upon it.

And what these sharks did to the snapper they *may* do to an unlucky human. At right: Arthur—a *Calypso* dummy. Equipped with backpack and fins, Arthur looked like a real diver, but during a series of tests sharks showed no interest in him until we put fish inside his wet suit. Now their interest was aroused. They began to circle him more rapidly, growing more and more excited as Arthur jiggled erratically at the end of his tether. One shark finally rushed Arthur, but veered off at the last moment. A second rush followed. Then a third shark wheeled and, with jaws agape, attacked the dummy, crunching down on one leg and tearing it off.

How Cone Snails Stab Prey

The deadly cone snail extends a hollow, venom-filled tooth from the end of its proboscis and shoots it into its quarry, where it remains. The small, barbed tooth, which looks like a miniature harpoon, is one of many radular teeth in the cone snail's radular sac. These harpoons have their origin in the file-like radular organ which is used by many snails to scrape algae off rocks. Only one tooth at a time is positioned at the tip of the proboscis, ready for use in defense or in capturing food. When a tooth is shot out, another radular tooth moves up to take its place. How the harpoon is actually shot out is not clearly understood—it may be with muscles in much the same way a sling-shot is used or through a high pressure system that rapidly forces the dart out. The venom is stored in a muscular, bulblike poison sac, which squeezes the toxin through a duct into the hollow tooth. In some species the venom is already in the tooth when it is fired and the snail can release the tooth entirely when it has found its target. In others the snail holds on to the tooth by means of a duct through which it pumps its poison into its prey. The venom paralyzes the cone snail's victim. Then the snail moves forward to engulf its immobile prey with an extensible and fleshy mouth. Because the snail's radula cannot scrape off tissue, a large meal must be partially digested in the pharynx outside the body.

There are many types of cone snails, most of which have beautiful shells. The Gloria Marus is an extremely valuable cone snail at times being appraised at over one thousand dollars. Because of their beauty, collectors look for them, but a number of people have received serious stings from cone snails in the South Pacific. These collectors, unaware

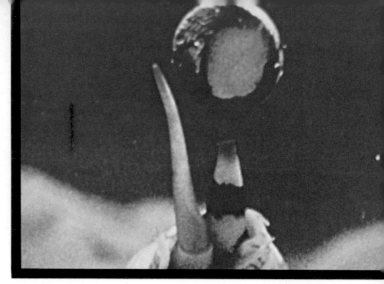

Preparing for attack. *The cone snail in the foreground prepares to attack a turban snail. The cone snail's proboscis is extended for the attack.*

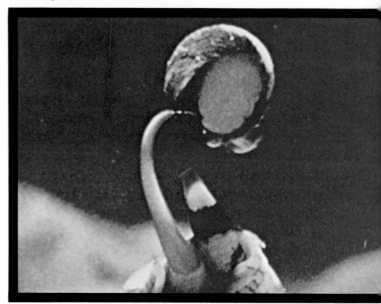

Successful attack. *When the victim's skin is punctured by the cone snail's tooth, venom has already begun to spread within its victim.*

of the deadly venom apparatus, have held cone snails in their hands or put them in their pockets. This has often caused the snail to react by releasing their weapon. The poison is reported to act on the nervous system and death in experimental animals may be from loss of respiration or heart failure.

Cone snail. *At right, an illustration of the stabbing apparatus of the cone snail, showing the poison sac, harpoonlike teeth, and the proboscis through which the hollow tooth is ejected.*

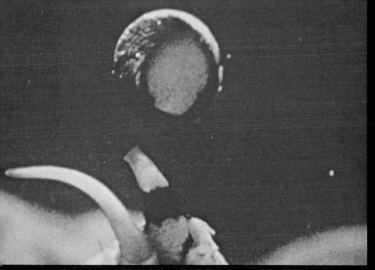

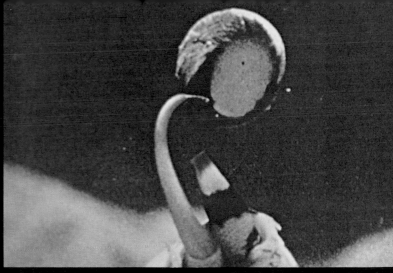

Proboscis in position. *As it moves to strike at the nearby snail, the cone snail waves its venomous tooth in the tip of its proboscis.*

Striking the victim. *The cone snail's proboscis curves around as the snail aims and lets fly its radular tooth. The victim snail is struck.*

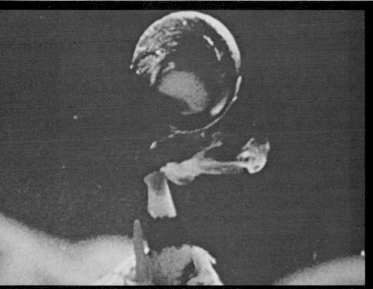

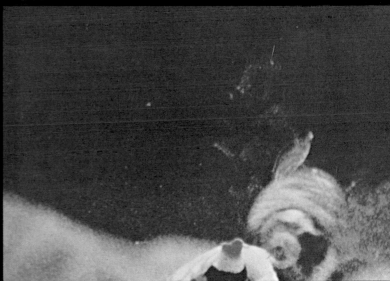

Victorious withdrawal. *The cloud of body fluids from the victim colors the water; the venom begins to affect the victim.*

To the victor the spoils. *In a few seconds the deadly venom has paralyzed the victim which will die within a few minutes. The snail will then eat.*

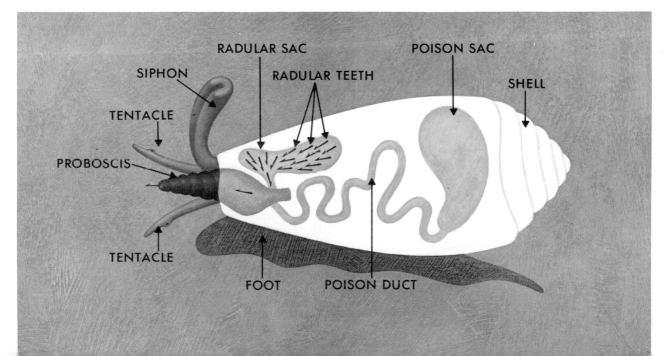

SIPHON

RADULAR SAC

RADULAR TEETH

POISON SAC

TENTACLE

SHELL

PROBOSCIS

TENTACLE

FOOT

POISON DUCT

Anemone—Passive Predator

Sea anemones, looking like flowers and growing attached to rocks and piles, are found in nearly all seas around the world, from shallow waters to depths of many thousands of feet. They prey on passing planktonic animals, worms, fish, crabs and even sea stars. When a tealia anemone feeds on tiny food particles, it expands its many delicate tentacles and pulls the food into its mouth. When a larger animal comes into contact with it, the anemone discharges the stinging nematocysts that line its tentacles. The nematocysts deliver a toxic substance, powerful enough to paralyze the animal so that it can then be eaten by the anemone. One of the only animals apparently immune from the anemone's stinging nematocysts is the little clownfish, a fish that frequently turns to the anemone for protection from predators. In return for this "favor," it is believed that the clownfish lures other creatures to the anemone, keeps it clean, and bites off tentacles which have become diseased. In general the anemone only releases the nematocysts when two conditions are met—the anemone must receive the scent of an animal as well as feel its touch. Without the stimulation of another animal's odor, the anemone's nematocysts discharge only in response to vigorous rubbing. When it senses danger, the anemone will contract its tentacles, sealing itself off from harm.

Tentacles. Rows of short, tapering tentacles (above) surround the oral disk of the tealia anemone. The petallike appearance of the tentacles belies the fact that they are studded with stinging cells.

The sting of anemone (right) is potent enough to kill a starfish. This tealia is almost finished with its meal, and only one arm of the starfish protrudes from the tealia's mouth.

Birds Feeding

Seabirds, flying high above the ocean's surface, can see for miles in every direction. They are alert for any disturbance indicating that small fish are being chased from below by larger predators. When attacked, the small fish are driven toward the ceiling of the sea, where they churn the surface waters in their desperate attempts to escape. The predators, in hot pursuit, may also break through the surface, increasing the turmoil. Shearwaters or other seabirds, seeing the splashing of the fish, fly over, knowing they may be able to find a meal. When the birds find a school of small fish trapped near the surface, they begin an orgy of feeding. Caught between two predators, the small fish are easy targets and both predators enjoy good hunting.

From their vantage point high above the

surface, the hungry birds get an overview of the desperate battle. They can anticipate the escape routes of the fish. They can also judge when the fish are nearest the surface and can be most easily captured. The confusion soon reaches a fever pitch, with predatory fish lunging upward to their intended victims, and the birds diving at them from above. The prey frantically zigzags and jumps to avoid being eaten. Then, as suddenly as it began, the battle ends. The sated predators from the deep leave, and the prey,

*These **sooty shearwaters** are feeding on a school of fish frightened to the surface by larger fish. Huge flocks of these birds, numbering 100,000 or more, often take off in the early morning and remain at sea all day in search of food.*

no longer threatened from below, returns to safer depths, out of the reach of the hungry birds. The waters soon become calm. Such dramatic feasts in the open sea take place every morning at dawn, and, to a lesser degree, late in the afternoon.

181

Chapter III. Color Me Invisible

Camouflage creates deceptive appearances. Animals deceive for two main purposes: avoiding predation and obtaining nourishment. Animals may use camouflage defensively; by blending in with their environ-

> "Some sea animals are quick change artists, transforming themselves in the wink of an eye. Others require a little more time—but still get the job done."

ment, they create an illusion of invisibility. As an offensive weapon, camouflage enables an animal to approach its prey undetected. With some animals camouflaging is a full-time occupation, with others a part-time one.

Probably the most widely used form of camouflage among fish is countershading, or obliterative shading. It renders a fish invisible to a potential predator or prey as the animal's color blends into the surrounding water and matches its reflection of light. The dorsal surface of these fish is dark so that when viewed from above it will match the deep blue water below. Conversely, the white or silvery belly of most fish renders them practically invisible from below by matching the light sea surface above. Disruptive coloration is another form of camouflage which is especially common among fish of coral reefs. It's common among many other marine animals too. This form is comparable to the crazy-quilt patterns painted on warships and military vehicles. Another form of camouflage is cryptic coloration. It is common among marine animals, especially those that dwell on the sea floor. To match a background many animals have developed the ability to change color or intensity of color. Some are quick-change

artists, transforming themselves in the wink of an eye; others require a little more time but still get the job done. Cryptic coloration often reaches a climax of beauty in the excitement of spawning.

Directive and deflective markings are another form of deception practiced by the animals of the sea. These markings usually draw attention to the least vulnerable part of the animal; or they may draw attention away from a particularly vulnerable area. Mimicry is another outstanding example of deception. An animal may seek to look like something else so it won't be recognized for what it really is. In some cases the animal mimics something inedible. In others, a predator resembles something harmless.

Many creatures that are brightly colored when in hand are virtually invisible in their normal habitat. Their reds or greens or blues have counterparts in their aquatic backgrounds, enabling them to blend with highly colored substrates. But there are some animals whose colors have not been so simply explained. In reality it may be impossible for us as humans to completely understand all the colors and markings of marine life. We can only see them through our own eyes and not through the eyes of a fish. We have a background and mental make-up entirely different from that of a snail or fish.

Protection by mimicry. As a means of surviving, some animals have developed a strong resemblance to another animal or inanimate object. Here, an amphipod crustacean (upper animal) has mimicked the appearance of a snail (lower animal). By doing so and by remaining close to the snail, the amphipod is frequently overlooked by predators. Mimicry is found also among some species of fish and other animals. In some cases, the imitator duplicates a poisonous animal, which is not usually preyed upon because of its toxicity. In other cases, it may mimic an inedible inanimate object.

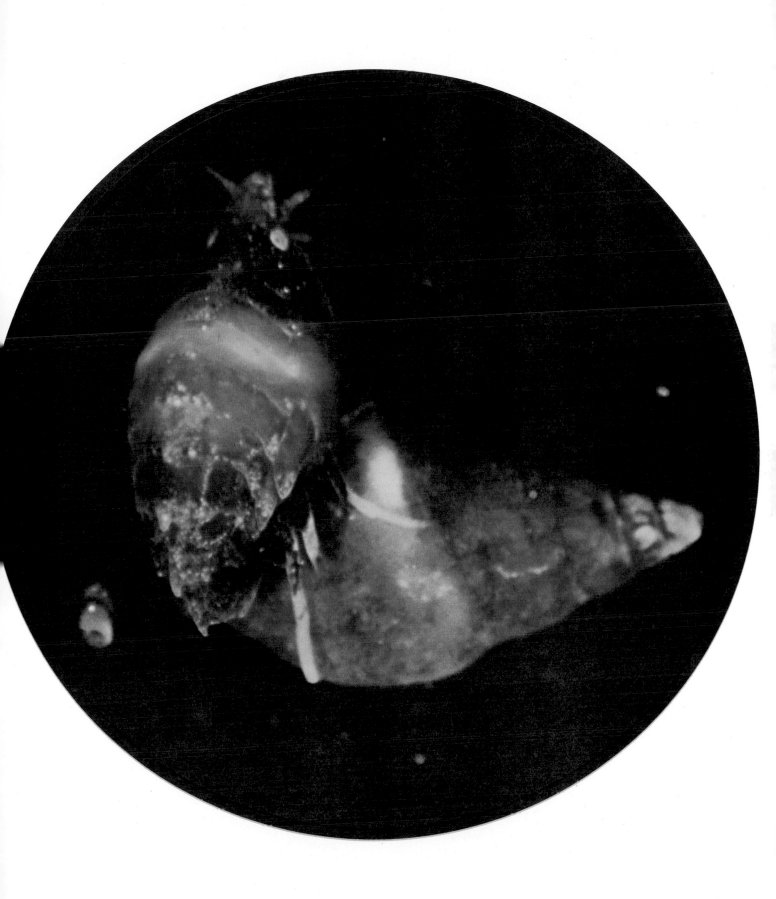

Fishing frog. Here we see cryptic coloration at its best. Deep in the Gulf of Aqaba, this fishing frog blends almost perfectly with a grayish sponge.

Israeli lizardfish. Lying in wait for its prey, the lizardfish depends on its cryptic coloration to avoid detection. It can strike with reptilian swiftness.

Find the Fish: Camouflage

The ability to change color helps some fishes avoid detection. They change color to match their backgrounds, blending so completely that even a person knowing of their presence may have trouble finding them. This concealing kind of pigmentation is called cryptic coloration. It enables many sea creatures to see but remain unseen. Thus they can seize their unsuspecting prey as it passes by, or they can remain invisible to their predators.

It is remarkable, considering the variety of colors fish can assume, that there actually are only four variables within their color repetory. They possess only three pigments —black, white and orange—and special re-

flective cells. It is these cells that permit fish to become an iridescent green or blue. Somehow they separate the spectral colors selectively reflecting some and absorbing others. Many fish have nervous control over the cells and can quickly alter the colors reflected. One of the most dramatic examples of this is the multicolored flashes that progress over the body of a spawning dolphin fish or mahimahi.

In the well lit levels of the sea an adjustable color system is essential. Bottom texture and colors vary from one place to another as does the color of the surrounding water. But as one goes deeper the colors of fish are less important and as a result become more limited. The most logical reason is that deeper

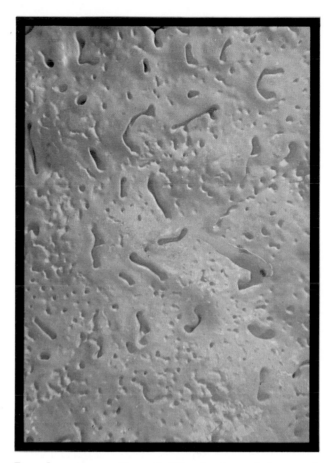

Camouflaged sculpin. *Blending with its pink background, this sculpin is barely visible even to a discerning eye.*

Royal gramma. *Even brilliant colors of the royal gramma can blend into its background. Here the fish remains close to a yellow orange sponge.*

in the sea colors cannot be seen and therefore are of little use. This is because a particular colored pigment only reflects that color of light and if that color of light is not present in the first place the pigment will have nothing to reflect and appear as a dark grey or black. Under these conditions a fish finding it advantageous to appear obscure or dark could be pigmented brown, black or even red, as many deep sea fish are, since those colors of light do not penetrate well.

In addition to cryptic coloration many bottom dwelling fish possess bizarre body extensions and protrusions to assist in camouflage. The scorpionfish family is a master at this; some produce fleshy extensions on the fins and head with hair-like projections all

along the lateral line. The advantages derived are most likely a breaking up of the fish's outline and an irregular texture not unlike the algae-covered bottom on which it rests.

Another camouflage technique is that of body posture. The animal may assume a position which tends to make it less apt to stand out. Pipefish are probably the best example. Their elongated shape, in a vertical position, closely resembles the marine grass in which they live. On the other hand, a horizontally swimming fish stands out dramatically against the background. Predatory trumpet-fish often position themselves parallel to a slender gorgonian in an attempt to remain unnoticed.

Find the Fish (continued)

Corambe. *At top, it is almost impossible to see the Corambe (right), a little nudibranch, nibbling on a lacy colony of bryozoans (left). The reason for the Corambe's resemblance to the bryozoans is hard to determine. The Corambe does not need to surprise the bryozoans, because these animals, like corals, live in rigid houses from which they cannot move. And, since it has few enemies, the Corambe doesn't require camouflage for protection.*

Klipfish. *The mottled appearance of this clinid fish (immediately above) blends with its multihued habitat. Additional protection comes from the spiny dorsal fin rays that disguise its form.*

Sea bass. *At right, this mottled sea bass lies on a reef and is not noticed by the smaller tang swimming past. When the bass moves to another spot on the reef, it adapts by changing color.*

Quick-Change Artist

The cephalopod cuttlefish, which is not a fish at all but a mollusk relative of the squid and octopus, is a master of disguise. It can change color in an instant or more gradually, whichever is more appropriate to circumstances. Unlike many other animals which can alter color only slowly through hormonal action, the cephalopods have nervous control over their skin pigment. Colored pigment is contained in cells called chromatophores and under stimulation they can expand or contract. What actually occurs is that many tiny muscles attached to the cells pull on the edges causing the cell to form a large flat plate making the pigment more apparent. Relaxation of the muscles causes the cell, and thus the pigment, to become concentrated into a dot and thus invisible. The varied colors that these remarkable animals can achieve result from a blending of pink, brown, blue, purple and black pigments. Coloration reflects the mood of the animal—white for fear, red for anger, multicolor for sexual display.

Chromatophores also can exist in fish but differ in the mechanism of pigment dispersion. For example, the melanophores, dark pigment cells, have many branching extensions into which pigment can be forced. The pigment granules move while the cell remains basically the same shape. By a similar controlling mechanism, orange and white pigments can greatly increase a fish's potential for coloration.

Cuttlefish above has become an indistinct gray, enabling it to blend into its background.

Cuttlefish at right displays large, black spots prior to change, which will fade the spots into vague blotches.

Deceptive Eyes

The eye of a fish, even though it may be a target for attacking predators, cannot be altered much to avoid its conspicuousness. Tissues surrounding the eye, and the iris itself may become pigmented to match the color of the body but there will always remain a jet black pupil. Because this black spot on the head is inevitable in all fish, many have evolved with color patterns that obscure the eye or detract from its prominence. For instance, a black bar on the fish that extends through the eye region makes the eye very difficult to distinguish and is surely advantageous to its owner. Another even more creative color pattern is the eye spot near the posterior part of the body seen on many butterfly fish, flatfish, wrasses, damsel fish, gobies, blennies and even on some rays. One example of the effectiveness of such an eye spot is in the misguided attacks of an Indo-Pacific blenny that preys on other larger fish by tearing off pieces of skin. It often goes for the eyes, but when confronting a butterfly fish it is directed to the eyespot at the rear of the body, which does much less harm to its victim.

Zanzibar butterfly fish. *(Above.) A distinctive directive mark calls attention of predators to less vulnerable parts of this fish. Black lines break up the animal's outline and give it added protection.*

Golden longnose butterfly fish. *(Right.) Not only does this fish have a directive mark at the rear end of its dorsal fin, but it also has stripes that help break up its outline and further confuse its enemies.*

Two-spot octopus. *(Upper right, page 191.) Eyespots on this octopus give the whole animal the appearance of being a huge head, which frightens off most predators. When it stretches itself out, it looks even more fearsome.*

Foureye butterfly fish. *(Lower right, page 191.) The false eye near the fish's tail draws the attention and perhaps the attack of predators away from the most vulnerable part of the fish—the head. This coloring, which confuses an attacker, is called a directive mark.*

190

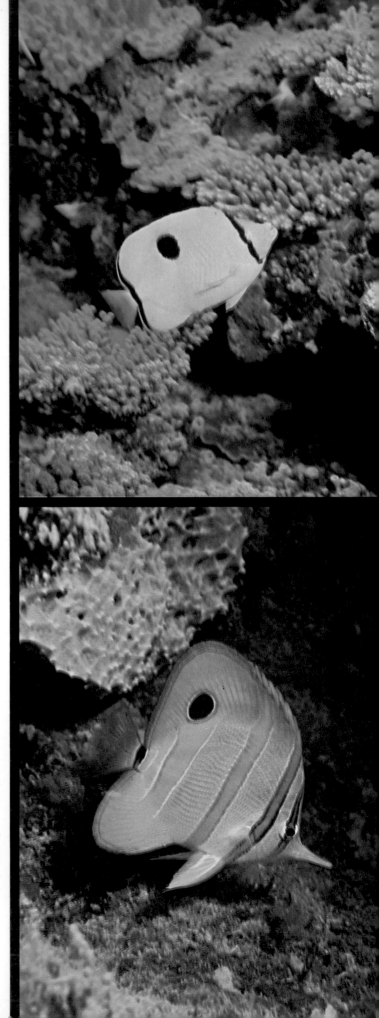

Moorish idol. Two nearly vertical black stripes deceive the eye and thereby make the actual form of this fish unrecognizable and difficult for a predator to attack at the proper angle.

Disruptive Coloration

Stripes, bars, and patches on a fish break up its outline so its fish form isn't obvious to an observer. To escape notice by a predator a potential prey must divert attention away from itself as a whole. Successful coloration and markings of the prey will create an optical illusion to the predator. A striking example of this is the reef fish which has a black band of color that extends through its eye. When a predator strikes, the attack is usually directed at the front of the fish with the eye as a point of focus. By obliterating that distinct feature with a black band the prey is given some degree of protection. The

Butterfly fish. *This fish is a good example of disruptive coloration. The black on its head blends into the dark background of the reef it swims in. The yellow blotches on its body break up the white areas.*

role of light and distance on a fish's appearance is another factor to consider in terms of coloration. To a diver looking at reef fish with artificial light and at close range the imposing light and dark markings are impossible to overlook. But under the naturally dim play of light objects a few yards away are often indistinguishable, and that same fish appears to vanish as its colors blend together in the distance. Another aspect to consider is the image seen by a predator. Studies have shown that many fish do not see as well as man and that some cannot even see different colors. With such handicaps a predator must be an amazingly efficient organism to be successful.

Stripes That Hide

To a schooling fish any markings that confuse a predator will increase its chance of survival. One effective means of confusion is having dark and light stripes which obliterate the distinct outline of each individual fish. The stripes of schooling fish are most often horizontal but a number of vertically striped species do exist. In a school of striped fish an attacking predator is faced with the problem of identifying an individual prey

Vertical stripes. In open-ocean fishes such as these jacks, bold bars on their sides may hide them. Although such stark markings may be highly visible close up, they fade into gray when viewed at a distance in the open sea, especially when the fish are schooling. The light and dark areas blend together when seen through the water and make the school a hazy mass to the observer.

and then pursuing that fish until its ultimate capture. All that is visible is a writhing mass of flashing lines, there is motion everywhere, no single fish can be differentiated from its neighbor. Hundreds of eyes and hundreds of lateral lines perceive the predators at every turn, all fish responding to it individually, none colliding. Some people have described a school of fish as a superorganism in which many individual units, all the same, function together for the good of the whole. In a sense this analogy is valid because many schooling fish could not survive alone if they were separated from the group. A school appears to be one large fluid unit, rather than hundreds of small, vulnerable animals.

In a school there are more individuals to sense a predator which may be some advantage to the group. The compact schools of some small fish may actually deter predators by appearing to be an object too large to attack. Some statistical research has shown that a school of fish in a given area will have fewer chance encounters with predators than if the individual fish were evenly dispersed. This means that a predator, assuming he could eat only so much at a time, would catch more fish over a period of time if they remained separate

There are a number of other advantages fish derive from living together in schools. More eyes and nostrils are available to seek out or detect food. Some fish change color as they begin to feed, possibly as a stimulus to others in the group to begin feeding. Many schooling fish feed on plankton, and their prey is far less likely to escape from a school than from a single fish. If a copepod darts away from one fish in a school, it will almost certainly find itself in the path of another. One further advantage for schooling fish is that no complex system of strenuous migration is necessary to bring the sexes together for reproduction. Breeding can take place merely by the simultaneous activities of the males and females. Inconveniences to schooling are rare but obvious—for example, when barracudas herd a school of sprats, feeding upon them for weeks until extinction.

Vertically-striped **damselfish,** *such as those above, are highly territorial, and at the slightest hint of danger will race to hide in the coral reefs they usually inhabit. As we can see in the bottom of the photograph, some of the fish have already chosen to conceal themselves in the coral. The large fish in the photo's center may be the reason the little damselfish are trying to hide.*

Grunts and jacks. *In the photograph covering the next two pages, the grunts, below, with their horizontal stripes and vivid coloration, remain close to the sun-dappled bottom where they can more easily blend in with the colors of the undersea terrain. The jacks, above them in midwater, are typically countershaded—darker above than below.*

Blending with Background Light

The silvery color of these anchovies, which are being chased by a yellow-tailed snapper, results from a reflecting layer in their scales. The layer is composed of iridocytes, opaque crystals of a waste product called guanine. These crystals reflect light in different ways, providing a silvery color at times and a white appearance at others. Additional layers of iridocytes and a mixture of iridocytes in the layers of normal pigment result in irides-

cence. The iridescent quality of butterfly wings probably results from a similar organization of pigment and reflecting material. How the light is actually reflected is not certain, but it may be that closely packed parallel layers of material allow certain wavelengths or colors to be reflected at a particular angle while others are absorbed. In any case, when light reflecting off fish which have iridescent coloring is close to the

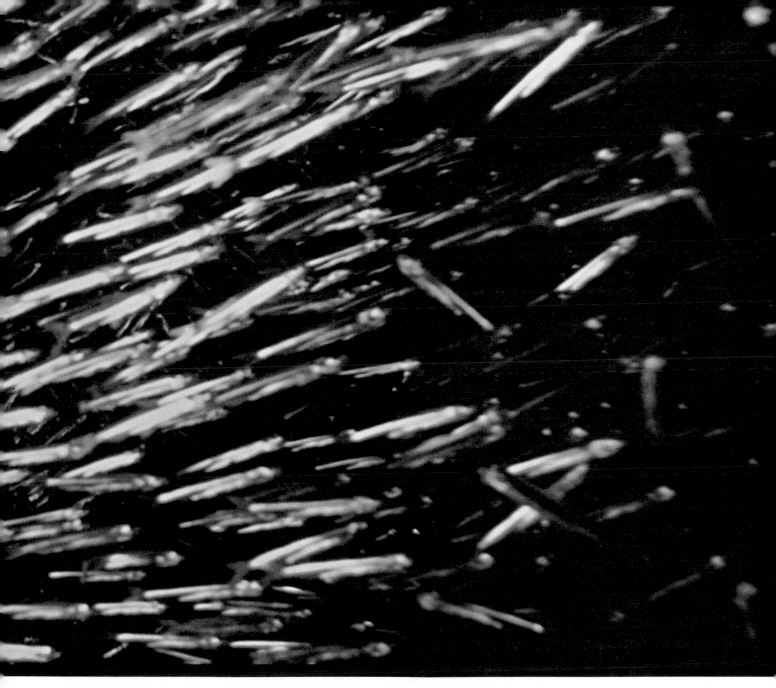

intensity of the natural background light, they become nearly invisible. But this yellowtail is close enough to see these anchovies. If it misses getting one, it will probably be because it couldn't zero in on a single fish since so many are present. The play of light from a group of these fish is a beautiful spectacle but trying to keep an eye on one individual in such a shimmering mass is almost impossible. This is the intended effect and is a distinct *advantage* in confusing an attacking predator. Experiments with

Yellowtail and anchovies. These anchovies, chased by a yellowtail, appear silver due to their reflecting layer of scales.

schooling fish in which a few were abnormally colored with a dye have shown that predators immediately select them as targets and are more successful at capturing them than normal fish. In other words, predators will weed out any abnormal fish making the species as a whole better able to survive. The "fittest," in this case, are those that blend with the school.

199

Cooperative Roundup

To a fisherman peering into the water almost all fish are invisible either due to the opacity of the water itself or the countershading of the fish. Because of this, his success may depend upon his ability to observe signs indicating the presence of fish. A school of fish feeding or being fed upon often disturbs the surface. Birds circle above and dive periodically amidst the animals scrambling for a meal. Their activity can be seen for consider-

able distances and act as a sign to fishermen. For instance, in this Mauritanian beach scene, man, bird and beast combine efforts to garner a harvest of mullets. These schooling mullets may have been chased inshore by other animals or may have been feeding there. Perhaps the seabirds above saw them first and inadvertently signalled the fishermen below. But when the men beat the water with long sticks dolphins drive the school of mullets into the nets of the fishermen and simultaneously feed on them. The

seabirds pick off pieces of fish left over by the dolphins. Soon the school is decimated.

There are other indirect methods of locating fish schools that to some extent compensate for the invisibility of fish to fishermen. Areas of dense plankton populations can be easily sighted by the color of the water and may indicate the presence of useful fish. At night the bioluminescence caused by schools as they swim through the water is another, if more limited, indication. In addition to these

Working together. In the photograph above, fishermen are rounding up a school of mullets. Sometimes circling birds give the fishermen a clue as to where to find a school of fish, and sometimes dolphins assist by herding the fish toward shore.

natural signals, man employs a host of scientific equipment to pinpoint fish far below the surface. Fishermen must depend on schooling behavior for large harvests of fish. Consequently the normally advantageous schooling way of life becomes a liability in terms of man's impact on the sea.

Chapter IV. Living in Armor

In the undersea world of predator and prey one of the best defenses is armor. And animals display a wide variety of armor, some of which man has been able to copy very effectively in his own weaponry.

But animals that adapt to life in armor face a major problem—they are cumbersome and slow-moving.

Animal armor comes in several styles. There are exoskeletons—skeletons that are external instead of internal as is man's. There are tubes, shells, and cases that are part of the animal. Exoskeletons are widespread even in the plant world. Diatoms, the basic ingredients in the ocean's intricate food webs, have silicate shells which protect them from chemical assault by the sea. Other microscopic marine plants—the flagellates—have shell-like plates protecting them. A number of seaweeds secrete and deposit on their exteriors a coating of lime that armors them.

In the animal world, the examples of exoskeletons are legion. Some of the one-celled protozoans, sponges, and corals—including the stinging corals, which are hydrozoans, and reef-building corals, which are madreporarians—have exoskeletons. The armor of crustaceans, including lobsters, crabs, shrimps, and barnacles, is familiar to many. And molluscs, like clams, scallops, mussels and oysters, are also well-protected.

Then there are the creatures that are born without armor but eventually live within cases, tubes, and tests of their own making. Some of armors start out as soft mucous secretions; these secretions combine with limy solutions and develop into tough outer coatings, which can ward off physical or chemical attack. Some other creatures that are born without armor inhabit the abandoned shells of other animals. Or these creatures make a safe haven in the substrates. They find protection as they gradually envelop themselves in the substrate.

In a class by themselves are certain fish and reptiles that are clad in a different kind of armor—scales, which can be thick and tough. Some fish, however, have evolved tough, scaleless skins.

Man's most obvious imitations of these animal armors are our military tanks. But there

> "Man's most obvious imitation of animal armor is the military tank. But there are other examples: ancient ironclads bristling with steel stakes, contemporary battleships, armored cars, air-raid shelters. And all buildings—a form of armor."

are other examples, like early ironclad naval vessels, which were armor-plated and often bristled with steel stakes. Contemporary warships have plates of the toughest steel. And armored cars, which tool about our city streets with bank funds, are covered with almost impervious plates. Air-raid shelters in warring countries are a form of armor. And all buildings in a sense are armor.

Dual armor. Protective armor comes in several forms. This pencil urchin, an echinoderm related to the sea star, carries stout clublike spines on its rugged calcareous outer shell (or test). Its spines are tough enough to ward off the attacks of many animals. And to repel the predators that manage to get through this part of its armor, the pencil urchin has a second line of defense—an inner test, which encloses its vital organs. Yet a few animals, like crabs, sea stars, and large fish, can penetrate the pencil urchin's dual armor and feed on it.

Heavy Armor

Molluscs owe much of their success to the heavy armor they carry about. Their ancestors probably possessed a simple horny covering for protection over 500 million years ago and subsequently gained the ability to impregnate this covering with calcium carbonate. This shell, along with a strong muscular foot, may have allowed snails, the largest and most successful group of molluscs, to exploit habitats too inhospitable for other animals. Fossil evidence indicates that much of mollusc evolution took place in the shore zone where an abundance of food and a variety of habitats existed. An impervious shell would have provided protection from both drying at low tide and abrasion and a strong muscular foot would have enabled them to hang on tightly to wave-swept rocks. One mollusc that successfully endures such a habitat is the limpet, a small animal which possesses a pyramid-shaped shell. The pointed shell probably reduces the unsettling effects of waves allowing the animal to graze on algae otherwise not utilized because of the environment in which they grow. Some of the primitive gastropods had coiled shells but originally they were coiled in one direction or plane, much like the shell of a chambered nautilus. Evolutionary modifications of this shell have made it more compact and better balanced and have resulted in the spiral conical configuration seen on many snails.

In addition to the protective coiled shell seen on most gastropods, a horny "trap door" or operculum further isolates the animal from outside. This durable shield may be any shape depending on the aperture of the shell. The queen conch has a curved narrow operculum that fits way back inside the shell while the turbo snail has a beautiful circular calcareous operculum often called the "cats-eye."

A hard shell is an effective protection, but these mollusks, however, are not invulnerable. A number of predators are successful at circumventing this deterrent. Bat rays possess grinding platelike teeth able to crunch the heaviest shelled clams; starfish extrude their stomachs to digest the protected snail.

Eggshell cowry. With its pure white, glossy exterior this cowry (opposite page, left) stands out clearly against almost any background.

The same cowry (opposite page, right) has covered its white shell with its black mantle. The mantle is a fleshy body wall that fits in the shell and can be extruded to camouflage it.

Triton. Right, it is large, heavy, and well-disguised by the growth of algae on its shell.

The same triton, below, withdrawn into its shell, displays his horny operculum (trap door), an additional protection from attackers.

Armor

The crustacea are a group of animals with a hard armor-like skeleton on the outside of their bodies. Physical protection is the most obvious advantage to having an exoskeleton because it permits the animal to lumber about in broad daylight, caring little for danger from others nearby. One disadvantage is the inability of a rigid outer covering to accommodate growth. As a result the vital protection must be cast off periodically and a new larger one constructed. During this time of transition the animal is extremely vulnerable and must remain hidden until the new shell hardens. An exoskeleton also

Pacific rock crab. This animal has an exoskeleton armor, as do most other crabs. Because of this hard, brittle exterior, the rock crab must shed its exoskeleton periodically to accommodate normal body growth.

limits freedom of movement. Any animal with a rigid covering will only be able to bend at its joints, so naturally its motion will be restricted and clumsy.

Some crabs find it advantageous to inhabit the empty shell of a snail rather than fabricate a complete covering for themselves.

These aptly named hermit crabs have a soft posterior that is curled to fit the spiral of their borrowed homes. They too must seek new protection as they grow and when moving to a new larger shell must expose their unprotected body to possible predation. Considering their fleshy abdomen they are surprisingly difficult to pull from a shell.

Hermit crab. *This crustacean has a soft body. To protect itself it adopts the abandoned shells of other sea creatures, and as it grows it searches for larger abandoned shells. It can be found in shallow waters around the world.*

Miniature Monster

The horseshoe crab pictured above is the epitome of an armored animal with its chitinous, rounded shell adorned with barnacles on its back and with a long, spikelike tail. This animal, although commonly called a crab, is not really a crustacean but is rather a member of the arachnid group, making it more closely allied to spiders than crabs. Unlike the crabs, this primitive arthropod has retained its tail, which it uses in righting itself when it is overturned. It is also a much slower and more awkward swimmer. These primitive creatures, whose form has remained unchanged for some 200 million years may burrow into sand for extra protection. The armored adults contain very little meat and are surely unpalatable to most predators but the young do succumb to some birds and fish. As the young grow older and larger they molt periodically. When they are full grown they keep their armored shell, as seen above.

Loggerhead Turtle

The body of this loggerhead turtle, like that of other sea turtles, is protected by the bony armor of its shell. The shell is made of two layers. The outer layer of plates is derived from reptilian scales and the inner layer has its origin in bony tissue. The points of junction for the first layer do not lie directly above those of the bony layer, and this gives the shell extra strength. Unlike the exoskeleton of crustacea, the turtle shell grows at the edges of individual plates of both layers, and this growth can be measured by rings or markings on the shell. The turtle continues to grow throughout its life, and its shell grows along with it. The heavy scales of the skin covering its flippers and head, plus the hard bone of its skull, afford additional protection. When attacking, the toothless sea turtles use their sharp-edged jawbones, which are powered by strong muscles. Their broad flippers enable them to move well in the sea where they have few enemies.

Pincushion and Box

For animals that are not built for speed, other defenses are necessary. Two families of fish have solved the problem in a similar way: their bodies are too hard for a predator to handle.

The spiny puffer's body is densely covered with short, sharp spines, which are actually modified scales. When it is threatened, the puffer can inflate itself with water to make itself more imposing and to erect its piercing spines. The puffer has another guarantee against would-be predators: it has poisonous flesh. The poison is a neurotoxic, affecting the victim's nervous system. In man it shuts down signals from the brain to the diaphragm, so breathing becomes impossible. Sea creatures die after consuming puffer.

The boxfish and its relatives, the cowfish and the trunkfish, have developed a hard shell-like outer covering that approximates an exoskeleton. Besides their outer armor, boxfishes and their relatives rely on a poison they secrete as an additional protection. The bright color of the boxfish may serve as a warning to predators to stay away from this crusty little fish.

Threatened puffer. In the face of danger, the spiny pufferfish, above, inflates itself with water and erects its sharp spines. Its pincushion appearance is an effective deterrent to most of the pufferfish's enemies. Unfortunately, it backfires when the predator is man, who finds the inflated bodies of pufferfish and their relatives attractive as lampshades or curiosities.

Immobile boxfish. The hard outer covering of the boxfish protects it against attack from larger predators. However tiny parasites are sometimes able to get under its skin. At the right, a little neon goby seems to be nipping the immobile boxfish, but may be picking a parasite from it. Boxfish are usually found in tropical waters, where they feed on some of the smaller invertebrates.

Spines That Protect

Some fish have sharp protective spines. These spines may occur almost anywhere on the body and in some cases occur everywhere on the body. Spines are either modified scales or spiny rays of the fins or bones that project out from the fish's body. A classic example of such modified scales is found all over the spiny puffer. Surgeonfish and tangs also possess scales modified as razor-sharp scalpels which evolved from bony ridge scales on the caudal peduncle. These lancelets are attached at the posterior end, projecting forward, and can be erected or depressed at will. Another example is the dagger on the tail of the stingray.

Spiny rays of the fins have also a dermal origin and possibly developed from some kind of scale. Supporting the idea of an evolutionary development toward spiny rays is the fact that almost none of the primitive fish possesses stiff rays while many of the more advanced fish do. These spines are of great importance for the defense of many fish, acting as instruments for the injection of poison, making the owner difficult to swallow or merely providing an unpleasant stingy surface to deter enemies. The development of elaborate spinous fins are a characteristic of the scorpionfish family and notorious in the lion and turkey fish, and in many other bottom-dwelling fish. The dorsal spines of the weever fish are actually used in offense

—it has been said to attack divers with its venomous weapons. Spiny rays can be found on the dorsal fin, the pectorals, the pelvics, and the ventral fin—or on all of these. In some cases the mere erection of these spines as a threat may deter a predator. Some reef fishes, like the triggerfish or the file fish, are able to erect their spines in such a way as to make them impossible to pull from a hole or crevice in which they are hiding.

Among the fishes with bony spinous projections the scorpionfish are probably the best known. Covering their heads are sharp bones that project outward, rendering them safe from most predators. The spines may help these bottom-dwelling, well-camouflaged creatures in breaking up the outline of the fish and making it look like another irregularity on the background. Probably the most common part of the fish's body where defensive bony projections have developed is the operculum or gill cover.

French angelfish. *These fish (left) have spines on their gill covers which they can fan out to fend off attackers. Other angelfish also have these cheek spines.*

The Achilles tang *(right) has a scalpel-sharp spine on both sides of its caudal peduncle, the portion of the body just in front of the tail. The spines are at the back end of a brightly colored heart-shaped patch. This flash of color probably serves as a warning to attackers.*

Hard to Swallow

The shapes of some animals make them difficult to swallow. Certain deep-bodied reef fish, for example, have flattened body forms which makes them impossible to consume in one mouthful. A predator such as the grouper, because it lacks the ripping and tearing teeth necessary for biting off portions of its prey, must be able to engulf its victim whole and then "chew" it up internally with its pharangeal teeth. This feeding method excludes any fish which cannot be taken in one mouthful.

As a result, deep-bodied fish such as the angelfish, butterflyfish, surgeonfish, tangs, filefish and triggerfish may receive a certain amount of protection from predators like the grouper. This body shape has other defensive advantages. A short deep-bodied fish can maneuver very easily among coral heads on a reef, possibly fooling predators. This body shape, coupled with their vivid coloration—surgeonfish range from yellow to purple, filefish are sometimes a deep brown with white and black spots—makes it easy for these creatures to conceal themselves among the multicolored coral.

Seahorse. The irregular shape of the seahorse (left) gives predators a difficult task. Since most animals take in their prey headfirst, an attempt to ingest a seahorse must include maneuvering it into the right position. This often proves to be a hard job. The seahorse also has a tough and leathery, though scaleless, skin to protect it.

The long-nosed butterflyfish (lower right) has a shape that a predator can't swallow easily. But at least two other factors also help it survive. Its disruptive coloration disguises its fish shape and makes it less readily recognizable as fair game. And its dorsal spines, erect when the fish is threatened, can stick in a predator's throat.

Trumpetfish. With its elongated shape and bony head, the trumpetfish (top right) is not likely to be taken whole by another fish. When a predator tries to swallow it headfirst, the trumpetfish resists and usually must be bitten into chunks to be ingested.

Chapter V. Strategic Withdrawals

We are all familiar with Goldsmith's lines: "For he who fights and runs away/May live to fight another day." This statement has truth for life in the sea as well as on land. An

> "For he who fights and runs away
> May live to fight another day."

animal outmatched in a fight is wise to withdraw if it can. Some creatures spend their lives hiding on the sea bottom or living inside a shell, so their very existence is a form of strategic withdrawal. Others, less hermitic, are candidates for attack from all sides, and when threatened, many prefer to leave the battleground. But how do animals escape especially if a predator continues to pursue them? Most simply, some escape by turning and fleeing, outdistancing or outmaneuvering their opponent. When we think of one animal outrunning another, we usually think they must have great speed and so it sometimes is. But many animals we consider to be incredibly slow-moving can move just fast enough to outrun an animal seeking to make a meal out of them. On land, a fugitive tries to escape in two dimensions. In the sea, an unpredictable three-dimensional sharp turn gives the advantage to the pursued over the pursuer, even if he is substantially slower.

> "Many animals we consider to be incredibly slow-moving can move fast enough to outrun an animal seeking to make a meal of them."

Some burst out of water in flight, returning every few seconds to scull with their tails.

Then they are off again. If the deck of a ship is not too far above the surface, a variety of sea animals is sometimes found there in the morning. Surprisingly, some creatures not even known to be "flyers" show up, thus giving us new insight to their ways and habits. Until the voyage of the *Kon Tiki*, authorities generally ignored reports that squids jetted themselves right out of the ocean. But with the evidence gained on this voyage, they began taking a closer look at these remarkable molluscs.

Before the development of radar, man-made vessels also could hide. If a vessel was losing a naval battle and a fogbank was nearby, it might make a run for the fog hoping to escape its enemy and defeat. Some animals like octopus, squid, or aplysia use a similar tactic when they are confronted by a superior foe. While they do not have fog to hide behind or in, they produce a substance that is equally effective. And, since they have control over its production, they can decide when it should be brought into play.

When threatened, some marine creatures which live in the substrate or among plants may duck into pockmarks in coral reefs, cavities in rocks, or other holes. Or they may bury themselves in the sand.

The animal that knows when to flee and when to stand and fight is usually successful in its continuing struggle for survival.

An octopus flees before the benign pursuit of a diver from the Calypso. Some maneuvers the octopus may use to escape danger are changing color to camouflage itself, retreating into a hole its pursuer can't get into, or shooting a cloud of dark ink around itself to obscure its actual position. Or the octopus may simply jet away as it is doing here.

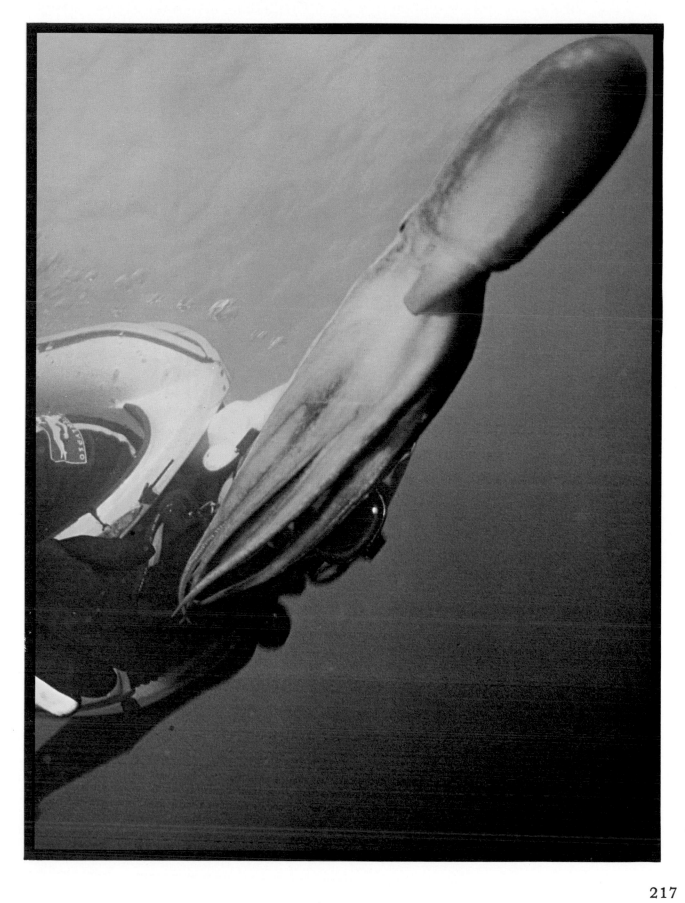

Jetting. A sea star (lower right) approaches two scallops. One scallop escapes the predatory sea star by bursting from the bottom in an explosion of sand.

Jet Propulsion

Encumbered by its shell, a scallop lives a fairly stationary life, spending most of its time on the bottom. But the shell is not an impregnable fortress, and when the scallop detects the approach of a predator, usually a sea star or an octopus, it flees, using its jet propulsion mechanism. Surprisingly enough, the direction of movement is forward with the open valves of the shells facing ahead. This is because as the shells clap shut water is forced out of small spaces on either side of the hinge. The jetting scallop's movement is jerky; and the scallop cannot escape a hungry octopus, but it can outdistance a sea star.

*Wild gyrations carry this pastel **nudibranch** away from the reaching arms of a sea star. Survival through open-ocean running.*

Gyrations

The nudibranch glides easily over the substrate, going about its daily routine. A predatory sea star, attracted by a chemical given off by this delicate animal, moves in for the kill. The sea star is a slow-moving predator, but its speed compares well with that of the nudibranch. Silently inching toward its intended victim, the sea star reaches out with one arm and touches the nudibranch. When contact is made, the nudibranch reacts. It gyrates wildly, trying to get off the sea floor and out of the grasp of its predator. When it finally begins to rise, it continues its gyrations until currents carry it to safety.

Flyingfish. *This fish has a remarkable way of escaping predators. With a burst of energy it flies out of the water and glides above the waves for as long as twelve seconds.*

Escape by Flight

In warm waters and far from land, this flyingfish, a juvenile in the Sargasso Sea, bursts through the surface of the sea when chased from below by predators. The attempt of flyingfish to escape to a safer locale brings them briefly into our world. At the approach of danger they accelerate toward the surface. When they gain enough speed to get airborne, they leave their predator behind. The fish skims over the wave tops, gliding first in one direction and then another. It may look down before landing; if it sees a predator still following, it sculls its tail in the water and scoots off in another direction.

*These two **mudskippers,** one partially hidden behind the plant, cling to a rock. If a predator attacks, they attempt escape by outrunning it or by jumping out of the water.*

The mudskipper is an unusual fish, since it is almost as much at home out of the water as it is in. It feeds on the mud flats exposed by receding tides. When it is alarmed, it bounds with great agility across the flat, skipping in a series of short, quick leaps. There it "walks" on modified pectoral and ventral fins, hopping about until the danger is passed. When it finds a crab's burrow or other crevice to hide in, it stays there until it feels safe.

When the mudskipper is threatened in deep water, it leaps out of the water. But it is not as skilled at staying airborne as the flying-fish, and it plops back in only a short distance from where it exited.

Jumping to Safety

Adult barracuda tend to be solitary hunters, while younger ones hunt in schools. These sleek, streamlined predators are machines of death for the smaller fish on which they prey. They have been observed herding fish into a compact group and then cutting a swath through them, snapping at and killing their prey without taking time to eat them. Then they turn around and eat at their leisure. Often the barracuda's attack is so swift that the herded fish cannot react to the marauders and become easy victims. But when the prey senses the coming attack, the schooled fish may leap out of the water in flight. In the air they are momentarily safe from the snapping, slashing teeth of the barracuda, but when they plop back in, they are again in danger. Flight reactions are not uncommon when schools of fish are attacked from below. At times it is possible to tell the path of a predator by the sequential jumping of fish in a line across the surface. Some of the

fish which become airborne most often are the halfbeaks, the needle fish and, of course, the flying fish. These are all sleek, surface dwelling fish which seek the nearest protective exit when threatened. Another marine animal prone to jumping is the squid. Squids have been reported to leave the water at dusk with amazing speed and hurtle through the air for a considerable distance.

There are other jumping animals which use their ability to jump for reasons other than

*The **barracuda**, sometimes called the wolf of the sea, may hunt in packs and herd its prey into a compact group.*

safety. The manta rays are marvelous jumpers, sometimes becoming so active that five or six may be in the air at the same time. Whales are another group that thrust their great hulks up out of the water then fall crashing back. Some think that such behavior is associated with a kind of mating ritual but these impressive feats do not always occur during the proper season.

223

Protection by Boring

Some animals spend nearly all their lives in hiding. Boring molluscs find safety by tunneling through mud, wood, and even rock.

Most clams burrow into sand or mud using only their soft foot as a digging implement. In Puget Sound along the coast of Washington the giant of all clams, the geoduck, may weigh twelve pounds and has a body long enough for its neck to reach the surface from a burrow as deep as four feet.

The largest and most efficient rock borers are the pholad clams. Near the end of their larval stage, the clams fasten themselves to a surface of heavy clay, sandstone, or limestone. They begin to make their burrow when their shells begin to form and harden. They move from side to side, or up and down, rasping the burrow face with the shells until a hole is gouged out. The movement and grinding continue, and soon the hole becomes a tunnel, which the clam increases in length and diameter to accommodate its own growth. In fact the mollusc be-

comes completely trapped within its chamber. The orientation of the shell is such that it can burrow forward but cannot reverse the direction. Thus the protective chamber becomes a prison with its only connection to the outside world being two siphons to the opening of the tunnel—one to bring in water and food and the other to expel water and body wastes. Many animals become so precisely adapted to their specific confined environment that they are unable to survive anywhere else. These burrowing clams are no exception, and possess reduced shells

Pholad clams. The shells of boring molluscs do not begin to form until the larvae settle on a surface they can bore into. In maturity the spherical shells are an efficient grinding apparatus, and the molluscs can scrape rock at the rate of an inch per year.

which are useless for protection because they are not large enough to cover the body. If released from their chambers they would be extremely vulnerable and unable to survive in the open. Boring molluscs are well-concealed within their tunnels. We often don't know of the presence of wood borers until the pier collapses.

225

Withdrawal. *Having sensed danger, the feather-dusters have withdrawn into their tubes. The anemone has folded its tentacles and is fully closed.*

Closing Up

The feather-duster worm and the tealia anemone are sessile animals that live fastened to the substrate. When they are threatened or disturbed they cannot move away easily, so they have devised other systems of "escape." The worm has very fragile gills, which it extends for feeding and respiration. If they are damaged, the worm might die. Fortunately the feather-duster secretes a rigid, fairly strong, tubular structure around itself. So, when the worm senses danger, it quickly withdraws into its tube, where it remains until the danger is passed.

The anemone does not construct a tube or

Feather-duster worms and tealia anemone.
On the left are three open feather dusters in feeding
position. On the right is an anemone.

other protective device, but it can withdraw into its own body cavity. When danger threatens, the anemone folds its tentacles toward its oral disc and then rolls them inside, until the sensitive tentacles are covered. After a short while, the anemone, tentatively begins to open up again, fully extending itself if the danger has passed.

"When the worm senses danger, it quickly withdraws into its tube, where it remains until the danger is passed. It has very fragile gills, and if they are damaged the animal will die."

227

Living In a Castle

Few divers exploring a coral garden are able to observe the full beauty of these delicate animals. Unless the flowerlike corals are extended to feed, which generally happens at night, they hide during the day inside the limestone castle they've built for themselves. Even more striking than the difference between the open and closed polyps is the dead, cleaned skeleton of a coral compared to the living specimen.

In a living coral the animal completely covers the hard calcium carbonate skeleton giving little indication of what intricacies lie below. A dead specimen whose tissues have been cleaned off shows the delicate skeletal

"Coral polyps are reasonably safe behind their hard walls. But a few predators can penetrate these defenses and feed on the flesh inside."

228

septa or projections that were secreted by the animal itself. These individual imprints are the easiest way to differentiate between species. However hard their protective skeletons, a number of reef fish, among which are the triggerfish and mainly all the parrotfish, have teeth or beaks hard and powerful enough to bite off chunks of coral—flesh and hardware—grind it in their throat, live on the flesh and eliminate fine sand.

When the little animals stretch out to feed, as most do at night, they use their tentacles,

Hidden. *Above left, the tiny animals of this small section of a brain coral colony have retreated to the safety of their limestone castle.*

Out to eat. *Above right, delicately colored brain coral polyps wave stinging tentacles to trap food. They retreat at the slightest disturbance.*

which are lined with stinging nematocysts, to entrap tiny planktonic animals. The poison of most corals is effective only against the minute forms that make up their diet and not against larger predators.

229

Tridacna clams. The shells of these three tridacnas are hidden amid a garden of sponges. In this picture their colorful mantles are exposed.

Sealed Tight

The tridacna is a fabled resident of the coral reefs of the Pacific. Legend holds that this large mollusc, sometimes four or five feet wide, is a killer clam, grabbing careless pearl divers between its two huge shells and holding them until they drown.

But this is all fantasy. The tridacna feeds largely on algal plants, which grow in its mantle between its shells. The shells are left agape to allow sunlight to reach the plants so that photosynthesis can occur. Tridacnas are extremely sensitive, and at even the slightest disturbance in the water they close their shells.

Bag of tricks. *A sea cucumber has some very effective defenses against attackers. Here it ejects
skeins of sticky white mucus to immobilize a starfish.*

Multiple Defenses

As they inch their way along the seabed, sea
cucumbers may seem extremely vulnerable.
In truth their defenses are formidable. Predators are discouraged by the poisonous skin
of some sea cucumbers. A cucumber that is
disturbed reacts by expelling water from its
body and contracting itself. A truly desperate sea cucumber resorts to a remarkable
defense: it turns itself inside out, spewing
out its insides—respiratory and reproduc-
tive organs and even its intestines—which
entangle its hapless attacker while the cucumber escapes. In about six weeks the
organs are regenerated.

Hiding In a Coral Head

Tiny, bright blue damselfish live in sunny tropical waters. Several hundred of them may inhabit a coral head, swimming just above it during the day and picking bits of food from the water. Generally the distance they swim from the coral head is indicative of the degree of danger around. The distant approach of a diver may cause them to move closer to their protective coral but not until the diver moves to within a certain critical distance will they instantly withdraw and hide in unison. It is remarkable to see how many of these apparently defenseless fish can completely disappear into a coral head out of reach of almost any predator. Because of their life style some species of damselfish never grow to any considerable size, therefore they can continue to live in their coral habitat. Individuals growing too large would be more easily captured by predators. Even if the coral head is removed from the water, the little fish does not leave it, but relies on its security to the last.

The arrow crab also seeks refuge by withdrawing into a protective shield of another animal. In this case the spines of a juvenile sea urchin provide a shelter under which the crab crawls. The hard shell of the crab's exoskeleton provides some protection for the tiny animal, but not enough to insure its survival. But the formidable spines that are part of the sea urchin's armor also aid the arrow crab.

Damselfish. *The small size of these apparently defenseless damselfish (above) enables them to live among protective coral branches.*

Spiny refuge. *An arrow crab (right) seeks protection among the sea urchin's spines.*

Covered with Sand

Many bottom-dwelling animals, like this angel shark and the partially buried ocipod crab, hide on the bottom to keep from being attacked. The angel shark, like the skates and rays, has a flattened body and broad pectoral fins. On the bottom, the shark flutters its pectorals to stir up sediments, which settle back on top of it and help to camouflage it. The sand breaks up the shark's outline, making it difficult for preda-

*An **angel shark** hides on the bottom, covered by a fine layer of bottom sediment. Note how well the brownish color of the shark's dorsal side blends with the color of the sand.*

tory animals to spy it. Hiding in this manner is carried to the extreme by members of the wrasse family. A number of these slender little fish are able to burrow in the sand, head first, and completely bury themselves when danger approaches. Some are reported to sleep through the night covered with sand.

This crab can quickly dig a burrow in the sandy bottom to conceal itself. It sifts the sand up and over its shell-covered body until only its eyes, situated at the tips of two stalks, protrude. Then it can watch what goes on around it without exposing itself to predators. The light-colored crab blends in especially well with the bottom, and until it moves out of its hiding place, it is extremely difficult to see. Relatives of this crab live just at the water's edge where waves sweep back and forth. With their antennae they filter plankton from the water as the waves recede. Below the sand they are virtually invisible. Their prominent antennae, however, give them away to birds searching for bits of food and fishermen in quest of bait. These telltale antennae may lead to the demise of the sand crab.

*This **crab** is an expert at excavation and can dig a burrow deep enough to conceal all but its eyes in a matter of seconds.*

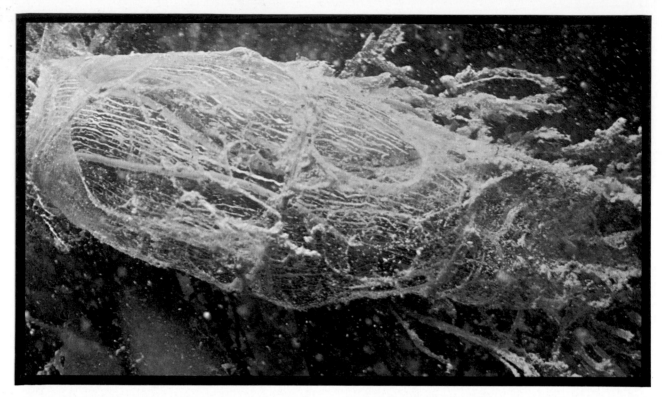

This **navanax** (above), found off the coast of California, is nearly hidden beneath strings of mucus it has secreted.

Navanax cocoon, below. When it has no further need of its protective cocoon, the navanax breaks open one end and eases out.

Mucous Shields

The brightly colored sea slug (navanax) is relatively safe from predation. It seems that only others of its own kind find it palatable. Perhaps one reason why predators leave it alone is the yellow fluid it gives off when disturbed. Another reason may be its odor, which is unpleasant to us and may also offend fish and other animals. The parent navanax imbeds its eggs in mucous coating to discourage predation. Adults also use mucous to form cocoons for their own protection.

Parrotfish are brightly colored residents of the coral reefs found in warm tropical seas. By day these fish graze on the reefs, biting off chunks of coral which they eat for the algae on and in them. By night some species sleep in a mucous envelope. This covering

Sleeping protected. Two queen parrotfish are caught napping in a crevice in the coral reef, surrounded by protective envelopes of mucus. Special glands in the fish's skin secrete the delicate-looking membrane.

completely surrounds the fish and may help protect it from its enemies. It may take the parrotfish as long as 30 minutes to secrete its covering. Some scientists think that the cocoon may act as a barrier to prevent the fish's odor from attracting a predator.

> "Some parrotfish sleep in a mucous envelope, which it may take them as long as 30 minutes to secrete. Although the covering looks fragile, it may take the fish nearly as long to break out of it as it took to build it."

237

Tunnels and Burrows

A blenny finds a hole in the substrate and simply moves in and adopts the territory around it as its own. The hole may be a crevice in coral, a tin can without a top, or even a broken bottle or other debris. Some blennies burrow into soft sand. Once it has staked out a claim to a piece of real estate, the blenny remains close to its home. It often rests in front of the opening, or just inside it, carefully watching for approaching predators, prey, or trespassers of its own species.

Garden eels dig into the sandy bottom with their tails, excavating long winding tunnels in which they hide at the first sign of danger. When they are not frightened, they reach partly out of the tunnel to feed on passing planktonic animals. They look like a garden of flexible question marks as they wave to and fro, facing the gentle, passing current.

On a trip to the Maldive Islands in the Indian Ocean, the crew of the *Calypso* observed these mysterious creatures. Areas of the bottom were carpeted with garden eels; but when a diver approached within 15 or 20 feet of them, the eels ducked into their burrows and were gone in the blink of an eye.

When the divers tried to remove one from its burrow to observe it, they discovered that the eels plug the entrance to their burrows with a mucus secreted from their skin. Finally some were captured and two were put into an aquarium on the bottom. Soon wrasses and triggerfish swarmed around the aquarium, trying to get at the helpless eels.

Garden eels. Above, garden eels in the sandy tunnels they dig with their tails. They withdraw at the slightest hint of danger.

Blenny. At right, a blenny peers out from its adopted home.

238

Chapter VI. Offensive Defenses

The drive for survival, which includes getting food and repelling predators, shapes defense and attack systems in most living things.

Poison is one of the defenses developed by a number of sea creatures. It may be administered by teeth, spines, beaks, or barbs. A

> "Man is not as well equipped physically for the survival race as are many other animals. His modern offensive and defensive devices are all inspired by nature: smokescreens, tear gas, poisons, knives and scythes, hammers and battering rams, stun guns and electric chairs, barbwire and burglar alarms, guns and slingshots. But with the invention of the nuclear bomb, man may have dissociated from nature."

few animals have electrical properties they use to stun prey or predator. Many crustaceans and some other animals have pinching claws to capture, crush, and rip food or to defend themselves against each other or other predators.

Whether poison is or is not involved, stabbing alone with sharp spines is often enough to discourage many predators from pursuit. Biting with teeth is one of the commonest defenses, especially among vertebrates. Some bite and hold, some slash as they bite, and some bite repeatedly; all cause blood to flow, thus damaging tissues and vital organs. Larger animals may use brute strength to club or ram an opponent. Sounds are some-

times used to frighten attackers away as much as to paralyze with fear a potential meal.

Some sea creatures change their shape to appear too large for the predator. And conversely a few can accommodate prey bigger than they are through expandable jaws and digestive tracts.

Man is not as well equipped physically for the survival race as are many other animals. With no claws, no fangs, a soft skin, and a modest running speed, he had to face constant hostility from the world and could only survive by fighting nature with all the tricks his brain and his hands enabled him to develop. Recently his conquest of energy sources and subsequent technology brought about the diversification of the weapons in his arsenal: they range from smokescreens, tear gas, and Mace to injections of poisons.

Modern man's devices for attack and defense seem more elaborate than animal systems. In reality, however, they are all inspired by nature, with the erratic exception of the nuclear bomb. One wonders if with this invention man has not dissociated himself from nature forever.

White tails. A hundred years ago, when man went out in small boats to harpoon whales, they feared the havoc that could be wrought by the tail of the whale. The flukes of the great whales could shatter a whaling boat. Whalers gave a name—"bobtailing"—to the whale's noisy habit of slapping their flukes on the water. In this picture a humpback whale's flukes flare upward as the whale begins to sound. Whales have been observed forming a protective circle around a wounded member of their pod. With their heads facing the injured whale and their flukes facing outward, they beat the sea furiously with their flukes to frighten their only traditional foes, orcas and white sharks. When used against predatory man, however, this usually effective defense backfires. Holding in place as they beat the water, the whales were shot by men from point-blank range.

A Spiny Defense

The numerous spines on this sculpin's back are an effective defense just as barbwire keeps intruders from a man's domain. In addition, the sculpin has spines on its cheeks, in its pectoral and pelvic fins, and in some species on its head. Any attempt against such a well-armed fish is disastrous for most predators. Mottled color and ragged appearance further help the sculpin evade trouble. But these defenses come into play only when the sculpin becomes afraid of being attacked. As a threat, the sculpin often spreads its broad, fanlike pectorals.

> "The sculpin has spines on its back, its cheeks, in its pectoral and pelvic fins, and in some species on its head. Any attempt against such a fish is disastrous for most predators."

Spines and Shape

Armed with venomous spines on its back, the scorpionfish is formidably defended. Since its ragged shape helps it blend into its usual habitat among rocky reefs, it is nearly invulnerable. Most members of the scorpionfish family have 12 or 13 spiny rays in their dorsal fins and most have venom in or at the foot of these rays. When a living creature brushes its body, the scorpionfish arches its back suddenly to stab the intruder.

Torpedo ray. *Lacking a stinging apparatus, the torpedo ray relies on its ability to produce an electric shock to stun its prey and defend itself.*

Electric Shock

The torpedo ray, also known as the electric ray, uses its unusual electric generator as an offense and a defense. Being a relatively sluggish swimmer, the torpedo could not capture its preys without its electric-shocking ability. It generates electricity in two large and two small organs on either side of the head; these organs are made up of many hexagonal cells filled with a jellylike substance. Large torpedoes can produce electric discharges of short duration but high voltage (220 volts). When repeated, the voltage decreases; organic batteries take some time to be chemically refueled.

Bat ray. The stinging spine at the end of the bat ray's tail is used as a defensive mechanism. And any animal hit by it will usually be killed.

Barbed Whiplike Tails

The bat ray has a short, stout, barbed stinging spine at the base of its long, whiplike tail. It is a strictly defensive weapon. The bat ray is an open-swimmer, but stabs an attacker by a flail of its tail in much the same manner as the bottom-dwelling rays do. To most animals hit by the barbed stinger the result is devastating. To man it is a serious but rarely fatal injury. A man hit by a bat ray's barb will feel intense pain accompanied by swelling, redness, and tenderness of the affected part plus swelling of the lymph gland serving that area. The pain and other symptoms may last several days.

Borrowed Poison

The delicate flowerlike sea anemones (center) and their relatives, the hydroids, all of which are really marine animals related to the jellyfish and corals, have stinging cells or nematocysts in their tentacles to stun their prey or protect them against predators. Some predators, however, like the nudibranchs (left and right) with white-tipped naked gills on their backs, are unaffected by stinging cells. In fact they eat them without destroying them, and use them for protection. These beautiful molluscs have the amazing ability to prevent the discharge of the stinging cells as they are consumed, as they pass into the digestive tract into and up a special canal, and are finally incorporated into their cerata or gills. It is known that the stinging cells have a projecting trigger that responds to touch and it seems logical that feeding actions would stimulate it. One possible explanation to the fact that they are not stimulated is that the nudibranch may inhibit the discharge of the nematocyst chemically, in much the same

way the clownfish gains immunity from the anemone. When predators attack nudibranchs, the ill effects of the stinging cells are passed on. Just how this is accomplished is not clear yet, but it has been proposed that cells discharge only in response to certain types of pressure or contact.

Another defensive adaptation of these naked snails is their coloration. Nudibranchs are among the most vividly colored animals in the sea possessing vibrant orange, blue, purple, yellow and red pigments. No predator could mistake any of them for a conventional prey. This is precisely the object of this eye-catching publicity for any animal that recognizes them (and it is assumed they do, having learned or having the ability to react instinctively) will not want to eat such a stinging snail. In contrast to camouflage, in other words, they are brightly colored, and advertise themselves as unpalatable.

Stunning beauty. *Sea anemones and hydroids stun everything that comes in contact with them—except the nudibranchs on the left and right of the photograph below.*

The Best Defense Is a Good Offense

Many of the more than 5000 varieties of nudibranchs display graceful crowns of soft respiratory organs (cerata) on their backs. Delicate, soft-bodied, and brightly colored, these delightful creatures would appear to be easy picking for any hungry predator. But these bizarre crawlers seem to flourish. The little animals secrete a mucus, which smells unpleasant to man and probably makes them unappetizing to fish and other predators.

When provoked, some species exude a strong acid and others discharge a powerful poison. The secretion of one species is reported to be fatal to fish and crustaceans. Another species has a specialized acid gland from which it releases a slimy sour secretion containing sulphuric acid. These are formidable reasons why nudibranchs make very little contribution to the ocean's food chain.

Some of the nudibranchs enjoy even further

Flashy warning. These closely related Southern California nudibranchs conspicuously display themselves, possibly warning they are not good to eat.

advantage. They are able to swim. They propel themselves by bending their bodies from side to side with head and tail almost touching. They can also beat the water with their cerata for additional speed. By moving up into the water they get away from any possible danger from an obstinate foe. Others are able to cast off parts of their bodies when they are under attack and get away. Later, these parts are regenerated.

Despite all these varied and formidable defenses nudibranchs are not immune to all predators. Some of them fall victim to parasitic worms and copepods and are eaten by some starfish.

These few predators aside, it seems that nudibranchs have enough going for them to be quite safe in their gaudy dress.

249

Poison Flesh

As if its covering of sharp spines were not enough to deter predators from the porcupine fish, many possess an additional defense: they are, when eaten, extremely poisonous. About one-half of the puffers and porcupine fish are reported to be toxic. The poisonous chemical is most concentrated in the ovaries or testes, the liver and the intestine. The muscles of the body are not poisonous but the skin is. The Japanese eat these fish but those who prepare these special dishes must undergo a period of training and must be licensed by the government. The Emperor is not allowed to eat puffer fish.

The symptoms of poisoning in mammals are an initial lethargy, weakness and loss of coordination. Gradually these problems of the nervous and muscular systems increase until death results. The action of the toxin is to disrupt the transmission of nerve impulses to muscles and certain centers of the brain.

Poison-Fang Blenny

This Pacific Ocean blenny has poisonous fangs, canine teeth in its lower jaw, that deter most predators. Groupers, large-mouthed fish that swallow their prey alive, have been seen spitting out a poison-fang blenny immediately after ingesting it. The blenny had probably bitten and poisoned the grouper. The blenny's bright coloration probably serves as a warning to would-be predators that the blenny is a bad risk. As with many poisonous animals the blenny has no fear of others larger than itself and even acts aggressively towards them. It can be assumed that this aggressive action has become a kind of behavioral warning to the fish's enemies. It threatens any intruder into its territory, including divers, with a series of short, jumplike strokes in the intruder's direction. There are, incidentally, two species of nonpoisonous blennies which look very much like the poison fang blenny, and they probably benefit greatly from this mimicry.

Suction-Cup Feet

A sea star, wrapped around blue mussel, pulls the shellfish's valves, or shells, apart by sticking its tube feet on either side and pulling in opposite directions. There has been considerable debate over the mechanism by which the sea star opens the valves of a mussel, clam or oyster. Some scientists claim that the sea star is stronger, thus forcing open the valves, or has a greater endurance and simply waits for the mussel to tire during a gradual pull. Others think that a narcotic chemical is secreted into the bivalve, enabling the predator to overcome its victim where it could not otherwise succeed. Whatever the method it uses, after opening the valves, the sea star everts its stomach *inside* the mussel's shells. Thus the mussel is eaten within its own shell. After consuming its prey, the sea star withdraws its stomach.

The attack. *The suction cup feet of this attacking sea star allow it to firmly grasp the shells of its victim and expose its flesh.*

Crown-of-Thorns

These tough-bodied sea stars crawl relentlessly over coral reefs and devour the living coral animals by everting their stomachs over them and digesting them on the spot. Unlike most sea stars, the crown-of-thorns has sharp, venom-bearing spines, which protect it against most predators. The spines consist of calcium carbonate and organic material and are held erect by muscles. Glandular tissue is reported to be contained within the spines themselves and can secrete a toxin into the water or into the tissue of a would-be predator. If a person is punctured by these spines, he will experience pain, redness, swelling, some muscle paralysis, nausea, and possibly vomiting. In addition to the venomous spines the crown-of-thorns has another poisonous chemical in the skin covering its entire body. The major enemy of the crown-of-thorns is another crawler, the triton snail, itself a large animal.

The venomous spines of the crown-of-thorns starfish are an effective deterrent to all animals except one—its only known predator—the triton snail.

Sea Urchin's Defenses

Sharp spines that sometimes transmit venom to the victim are the first line of defense of most sea urchins, which are relatives of the sea stars. Besides the spines, sea urchins have *pedicellariae,* three-part jaws on stalks. In venom-bearing species, the pedicellariae transmit the poison. Yet another defense sea urchins have is the shell, or test, in which spines and pedicellariae originate. Like the spines, the test is made of calcium carbonate and is hard but brittle. When attacked, a sea urchin brings its spines into play, waving them about and aiming their sharp tips at its attacker.

Some sea urchins, like this white-tipped one, have added an additional defense. As we see here, they are able to pick up debris from the bottom they dwell on; moving these loose fragments from one spine tip to the next, they finally place the debris on top of themselves. One reason for adopting this cover may be for protection from sunlight. The debris may also serve as camouflage. Thus,

an urchin living on tropical turtle-grass flats may be covered with bits of turtle grass. If it lies on a pebble-strewn bottom, as shown here, it will disguise itself with pebbles.

Sea urchins. *The sea urchins in the photographs below and to the right are perfectly fit to take care of themselves. Some species have movable spines, and in addition some have a structure at the base of these spines that give off a poison. Other species, like the one below, are capable of covering themselves as protection from harsh sunlight, and as a means of camouflage. Many sea urchins live on rocks in shallow water, and those that live deeper down tend to live in groups. They range in size from one-half inch to ten inches.*

Pistol shrimp. *The pistol shrimp above uses its large claw to create a snapping sound which stuns prey, startles predators.*

Creating a Loud Noise

Loud sounds frighten many animals, and some sea creatures use noises as a defense. For example, pistol shrimp use their large claws to produce the sharp snapping sound that gives them their name. The sound is so loud it can be heard easily by beachcombers or by oyster fishermen, as these snapping shrimp also live under clumps of oysters. But its defense also works against the shrimp; it can give away its presence when it might otherwise be unnoticed. Mediterranean fishermen often find distant rocky shallows out of sight of land by listening to the snapping shrimps.

Crab. *Protected by its exoskeleton, this tropical crab remains unconcerned by the presence of an underwater photographer.*

King Crab

The hard outer shell, or exoskeleton, of this tropical king crab is its principal defense. Its armor, coupled with its large size, makes it a tough consumer product. The pincers on the crabs' long, spidery legs can also give an attacker a good nip. Their cousins, the northern king crabs in Alaska, grow to 20 pounds and may be five feet wide; their size alone discourages many predators. The marine growth that accumulates on their 10- to 12-inch body shells or carapaces, acts as camouflage. When approached, most crabs thrust their arms upward brandishing open claws in a threatening manner.

257

Chapter VII. Fighting for Territory and Sex

When combat takes place between members of the same species, it is almost always the result of competition for territory or a mate. Initially, one animal may assume control of a territory by driving off another occupant or by just moving into an unoccupied area. If challenged for control of its home ground, the animal will defend it vigorously. Similarly, two males competing for the favors of a single female may engage in ferocious ritualistic combat for the prospective mate. In both cases, possession is the motivation.

Open-ocean species are generally not territorial, but inshore animals frequently are. For territorial animals, to own a home, unencroached upon by others, is critical to their survival both as individuals and as a species. Its territory gives an animal a place to hide from predators. It also provides a place to feed without the competition that could lead to overgrazing and subsequent shortage of food. Ownership of territory also plays a major role in the sex lives of some animals. The territory provides a place to mate and lay eggs or give birth to offspring.

Fighting for the right to mate is to the advantage of the whole species, since the strong characteristics of the victor are passed along to the offspring. The loser, usually not as fit as the winner, is prevented from passing along his weaker traits.

Although combat for sex and territory seem dangerous to the participants, injuries are usually not very serious and rarely fatal. Animals generally do not fight among themselves the same way they fight an enemy. In fighting members of their own species, they meet in ritualized combat, involving the use of threat displays and other noninjurious means to defeat a challenger. The weapons they use against predators and prey (such as teeth, claws, and poisons) are rarely or partly used against each other. These weapons are intended for killing, and kill is not in the interest of the species. So they resort to violent, but usually harmless, jousting. Though there are notorious exceptions, such as the deadly fights of octopus for a shelter, a growl, grunt or squeak may be

> "The size of participants is not indicative of their ferocity. The smallest creatures fight with at least the same resolve as the largest."

enough to drive an opponent away. Some fish beat at each other with their tail fins. The pressure wave they set up is indicative of the strength of the fish. Often, though, more subtle means are employed to show superiority. Sometimes opening the mouth wide is enough of a threat to frighten off a challenger. Color changes, ritual movements, or flaring of the fins may do the job too. The size of the participants is important in the intimidation game, but is not indicative of their ferocity. The smallest creatures fight with the same resolve as the largest until one finally emerges victorious and allows the loser to leave and fight another year.

*The **jawfish** is extremely territorial and does not like to wander far from the safety of its burrow. At the approach of another jawfish, it quickly darts out of its burrow and rushes toward the intruder as though to strike it. But it seldom hits its adversary; it returns to its burrow nearly as fast as it left, diving back inside when the confrontation is over. This fish is named for its highly eversible jaws, which it opens wide to threaten an intruder. An open mouth probably makes the fish appear larger than it is. As in most battles between animals of the same species, neither is harmed. Locking jaws is the most violent act of warring jawfish.*

Mantis shrimp. This animal will normally fight over territory, spreading its armored tail over the entrance to its lair. This usually deters any challengers.

Physical Combat

Most animals end a battle with others of their own kind before coming to blows. Mantis shrimp live in cavities in rocks, corals, sponges, or in empty shells, while some larger species prefer to drill their own cylindrical home in the sediments. Their fighting is generally over occupancy of a cavity. Combat between mantis shrimps follows a strict ritual. The challenger approaches boldly, head high, antennae in motion, eyes fixed on the goal. The occupying mantis spreads its armored tail to block the entrance to the cavity. This action is often enough to discourage the intruder.

260

Eel blind. To study and photograph a colony of garden eels in the Red Sea, a small blind was erected. Divers behind the structure did not disturb the animals.

Garden Eels

Colonies of garden eels live in individual burrows dug in the ocean floor. So timid of exposure is the garden eel that none has ever been seen wholly extended by uncamouflaged divers. There appears to be a rigid social order in garden eel colonies, with dominant males, harems, and firm property lines. A challenger from an adjoining territory is met with a ritual display. Stretching from the burrow, the eel performs snakelike undulations, ripples its dorsal fin, and turns its profile and then its back to the attacker. If the challenger persists, the eels will square off, snout to snout, and strike at each other.

Lobsters in Combat

Atlantic lobsters in combat use their large pincers. Here a pair face off and, unlike many other creatures in the sea, actually do bodily harm to each other. In this fight for territory, one lobster has succeeded in literally disarming the other. The loser could be killed and eaten by the cannibalistic opponent. Or it could limp off to regenerate a new claw to replace the old one. Regeneration to full size may take several moults over a two-year

period. Atlantic lobsters usually have their hard exoskeleton to protect them against other lobsters or other marauders, and if necessary, they can shed their claws to escape.

Lobsters are most vulnerable immediately after they have moulted. Males and females alike are subject to harassment and dismemberment by their own kind and others when their new shells are no stronger than wet paper. In the several days after they shed their old shell and before their new ones have hardened, they spend much of their

time in hiding. During this time they have their greatest need for calcium carbonate, which provides the hard substance of the shell. The lobster's new shell begins forming under the old one and is principally composed of soft proteinaceous material. This shell swells when the old one is shed and provides the animal a little room for growth. The action of sea water upon the protein-aceous material and the addition of calcium carbonate to it, causes the shell to harden. The lobster's need and craving for lime or

Battling lobsters. *The loser of this battle may lose anything from a claw to its life, depending on the cannibalistic nature of the victor.*

calcium carbonate is great, and it is not un-common for a lobster to consume its cast-off exoskeleton immediately after shedding it. It is the calcium carbonate deficiency that has made lobster mariculture difficult. Many attempts have been made to raise these suc-culent animals in captivity, but in the crowded tanks fighting is constant.

Fearless Guardians

Wrecks and coral or rock reefs are favorite haunts of groupers. When they find a cavern that suits their needs, they stake it out and call it home. They are extremely defensive of their home territory and do not allow passersby to trespass, regardless of their size. This pair was guarding their territory when its boundary was violated by the photographer. Feeling threatened, the male was openly aggressive and rushed the diver. It hovered brazenly before the camera, its mouth opened wide in characteristic threat display, until the diver backed down. Its partner hovered beneath and behind it, being slightly more cautious.

The brilliant orange garibaldi lives on the rocky shorelines of southern California and Baja California. It lives among the great kelp forests there and defends its territory year round. During July and August when the females are ready to spawn, the males become more defensive than normal about their home site. They fastidiously clean around a clump of red algae to prepare it for the female to deposit her eggs and courageously attack any animal who moves in with intent of taking it away.

Anyone who has observed this fish underwater, whether as a diver or from a glass bottom boat, has surely marvelled at its brilliant color, particularly when compared with its normally drab neighbors. Obviously, camouflage is not important to this fish. It has been postulated that such vivid colors may serve as a means of identification to others, which would supposedly deter them from approaching this pugnacious fish.

Pictured above are **two groupers,** *whose home territory is under threat. Being highly territorial animals, groupers fear little. By opening its mouth wide, the larger grouper displays its characteristic threat.*

This **garibaldi** *hovers alertly near its nest, ready to defend its territory against intruders, no matter what their size. In captivity, the little animal turns neurotically cannibalistic.*

Kissing Snappers

As these gray snappers square off in ritualized combat, they seem almost ready to kiss. In fact, the fish are trying to make themselves appear larger by opening their mouths as wide as possible. A confrontation generally does not go beyond this stage. For many reef fish, disputes arise over the control of territory, but conclusions are difficult to draw, as fish behavior is disturbed by the intrusion of an observer, and fish do not conduct themselves normally when they are in captivity.

Such bloodless confrontations are purely instinctive reactions, not a form of learned behavior. To prove this, experiments have been undertaken in which two animals were reared separately and in isolation from others of their species. When the isolated fish were introduced to another of the same species, they exhibited the inherent ritualized behavior seen in their normally reared relatives.

Aggressive behavior towards a member of another species may differ drastically from that towards a member of the same species. In attacking a nonrelated intruder a fish may actually use its teeth or whatever defenses are available to do physical harm. This difference shows that the non-destructive attacks or ritualized threats are really a form of communication. There are many advantages to having rules that enable opponents to threaten, attack and win or lose.

Threat display. *Despite looks to the contrary, these snappers are not ready to kiss. Their ritualized pose is a threat to one-another, probably over territorial rights.*

The victor could remain a productive individual unharmed by battle and the loser could eventually succeed as others became weak or died. The population as a whole would be able to keep its numbers up and remain viable. The field of international relations is an obvious analogy.

267

Fighting for Social Status

Battles between elephant seal bulls often occur during mating season, and are usually intended to determine the social order within the herd. Young bulls, not yet sexually mature, engage in mock bouts, rehearsing for adult combat. Before body contact is made, the bulls threaten each other with an inflated snout, a raised stance, abrupt aggressive movements and alarmingly boisterous bellows. When one of the opponents is obviously weaker, it will retreat before blood is shed. Only one in 60 of these confrontations gets beyond the threat stage. But if the males are comparable in size, weight and aggressiveness, threats are ignored, the animals square off and the fighting begins.

Nearly all fights occur on land, but the combatants may move off the beach and into the water. Land-based fights are generally short, lasting five minutes or less. But they may battle for 45 minutes.

The two behemoths stand chest to chest, feinting and faking, waiting for an opening to fight. Finally, with a fast and powerful blow, one strikes at the neck of its opponent. The attacker's head slashes downward and its sharp teeth rake the opponent's flesh.

When one has had enough, it backs away, conceding defeat. The loser has not necessarily received the worst of the battle, but for some reason it chooses not to hold its ground. The bleeding is profuse and the

Elephant seals. In the bloody scene above, two elephant seal bulls are locked in combat. Battles of this sort sometimes occur during the mating season, usually on land. The fighting is brutal but short lived, and although the wounds inflicted are deep, elephant seals heal rapidly.

wounds are deep, but they heal quickly. Fortunately the neck and chest of the elephant seal can stand up to this rugged treatment and have a horny layer of tissue to help protect the animals from serious injury.

Chapter VIII. Ancient Animosities

Relationships among people and other animals take many forms. Some species get along with each other. There are certain creatures who have so little to do with one another's lives that theirs must be termed a non-relationship. There is, for instance, the case of the anemone and the turban snail. They simply do not care about each other. They do not look to each other for food or shelter; nor do they crave the same food or shelter. The anemone eats microscopic

> "There are a few species which seem to have a natural animosity toward other species. These antagonisms give the appearance of being fraught with deep emotions of fear and hate."

plankton, and the snail's diet consists mostly of fine algae. Neither animal is interested in the kelp it rests on. There are many relationships like this one in the sea in which two or more animals can live in close proximity with one having no bearing on the lives of the others.

Some animals even help each other. But there are a few species which seem to have a natural animosity toward other particular species. Such are cats and dogs, cobras and mongooses, and in the ocean, sperm whale and giant squid. These antagonisms give the appearance of being fraught with deep emotions of fear and hate. Whether they actually are is highly speculative.

Why and how do these traditional animosities arise between species When two species don't get along, their aggressive behavior could be emotional or calculated. Perhaps hundreds of thousands of generations ago the members of two species may have com-

peted for the same habitat, territory, food, or ecological niche. In competing, they may have resorted to combat. And perhaps this combativeness has continued as one species faces the other, even many generations after one has adapted to a different niche and the two are no longer competitive. Octopus-lobster; moray-octopus; lobster-moray are three couples with such built-in animosities. Mediterranean fishermen tell stories of traps that they pull back to their boats, sometimes containing an octopus, a lobster, and a moray eel: the three retreat to the three corners of the trap as far from the others as possible, because they know the first one to attack will be immediately killed by the third party.

Hard-shelled snails have enemies that are able to overcome their rugged defenses. At right a large tulip snail is devouring a small helmet snail. The larger animal works on the trapdoor (operculum) of the smaller, gains entry to the shell, and consumes the soft flesh. Many snails are active predators on creatures they are able to catch in a slow-motion race; others consume marine algae.

Anemone and turban snail. They inhabit the same environment, but live in different worlds and pay no attention to each other.

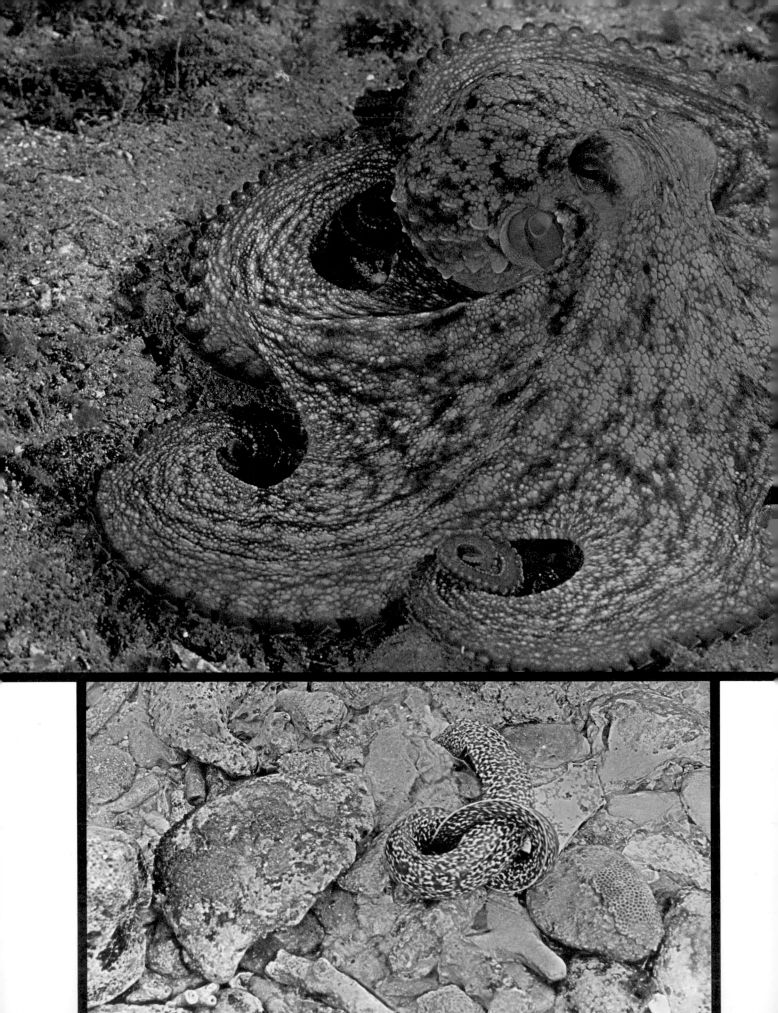

Moray Eel and Octopus

When octopus meets moray, a struggle to the death ensues between these mortal enemies. Morays usually prey on octopods at night. When a moray smells an octopus, the eel slips in and out of holes throughout the area, zeroing in on its quarry largely by scent (the moray's eyesight is not particularly good, especially when feeding at night). When it finds the octopus, it clamps its jaws on it and the struggle begins. The octopus

reacts immediately by wrapping its tentacles around the moray's head. The moray then ties its body into an overhand knot by looping its tail around and over itself. The eel withdraws through the knot sliding out of the octopus's grip while retaining a jawhold on it. Quickly, then, the moray can start gulping his prey.

The viselike grip of a moray, reported by divers who have been bitten, is almost impossible to break until the eel decides to give up. The teeth of morays are particularly adapted for grasping prey. Projecting from both the upper and lower jaws, and even the

> "The octopus reacts by wrapping its tentacles around the moray's head ... The moray ties its body into an overhand knot ... Then the eel withdraws through the knot, sliding out of the octopus's grip while retaining a jawhold on it."

roof of the mouth, are teeth that are sharp and slender like needles that are essential for any fish which must hang on to an octopus.

If the moray is not successful in its first couple of attacks on an octopus it will very likely go without a meal. This is because the ink of the octopus can act in two ways to assist its retreat. Film footage, has revealed that the first advantage to the octopus is that a cloud of ink presents a false object for the eel to strike at. The second asset lies in the fact that chemicals in the ink temporarily deaden the sense of smell in the eel.

Making a knot. Left, a spotted moray on a rocky beach ties itself into a knot. This is similar to its reaction to an octopus's grip.

An octopus (above) watches warily; it has changed its color to match its background and thus escape detection.

Penguins and Skuas

In the southernmost parts of the Southern Hemisphere, where the Adelie penguins pictured here normally nest, there is a species of predatory oceanic bird called the skua. These birds have characteristics of both the gull and the hawk. They are similar to the gull in size, habitat, and food preferences. And they have the hooked beak, the predatory habits, and the general appearance of the hawk. Skuas are related to gulls and to a group of seabirds called jaegers, a word which means

"hunter" in German. The term could well be applied to the skua as well. Twice a year, the skuas prey especially on the Adelie penguin. Normally the skuas are fish eaters, and they dive into the water to capture fish. But when the Adelie penguins lay their eggs, skuas returning from their fishing waters often seize and devour some of the eggs that are unattended. Not long after, when the chicks have hatched, skuas again prey on the weakest of the baby birds, ripping the helpless chicks with their hooked beaks. The baby penguins are defended by adults who very

courageously and very successfully chase the skuas. The skuas are frightfully obstinate; they remain nearby waiting for any young left unattended.

In the northern parts of the Northern Hemisphere, where skuas also occur, these hunters prey on fish and rarely on the arctic birds, such as gulls and terns that may come near them.

Skuas seem to have an underserved reputation for being a vicious predator. They are a predator no different than any other and

Adelie penguins. *Twice a year these penguins are threatened by a species of gull-like bird called the skua. They are ill-fitted to defend themselves.*

actually, in the long run, benefit the penguin population as a whole. Recent observations prove that most penguin eggs taken by the skuas were either unfertilized or dead before hatching and that their role in the rookery is more that of a scavenger than of a predator. They also might remove the behaviorally or physically less fit, indirectly making the species stronger.

Sea Star and Octopus

After being dropped in front of an octopus home, this sea star is attempting a getaway. Although the sea star and octopus may not be ancient enemies, it is obvious that their relationship is not pleasant, at least for the sea star. Sea stars are not noted for their speed, which is usually measured in inches per hour. The octopus is not known to prey upon sea stars, and this drama is typical of the perversion of instinct and behavior that man brings about.

A / Recognition. *The sea star rapidly flips over to right itself. The octopus watches.*

B / Sea star retreat. *The sea star begins to move away under the gaze of the octopus.*

C / The octopus moves. *The sea star stretches its tube feet to speed away from danger. Then the octopus moves into action. A small blenny watches.*

D / The chase continues. *As the octopus moves closer to the sea star, the blenny stays near, apparently an interested witness.*

E / Capture. *The sea star is unable to make good its escape. The octopus has covered it with a tentacle and will pull it back to its den. But what for?*

▲ A   ▼ B

Urchins and Their Enemies

Despite their impressive defensive equipment, sea urchins have many enemies. The most dramatic example of sea urchin predation is found in the feeding behavior of the queen triggerfish. The diadema sea urchin has an interesting defense system which must be overcome by the triggerfish. When any point on the urchin is touched, it directs all its long thin, highly mobile spines toward that point of contact. This makes it almost impossible for a predator to penetrate the spiny coat to attack the more vulnerable shell. The triggerfish cannot attack the dorsal surface and must, instead, focus its attention to the ventral or undersurface where the spines are very short. The problem for the fish is how to get to this vulnerable side. The solution lies in the queen triggerfish's ability to grasp the longest spines of the urchin in its teeth, swim up off the bottom, then let go. As the hold is released, the urchin simply falls to the bottom, usually ventral side down. The triggerfish is persistent, however, and continues until the urchin happens to land on its side. As soon as this occurs, the triggerfish is able to whirl around and attack the urchin's vulnerable underside. The short spines are not enough of a deterrent to prevent the sharp protruding teeth of the triggerfish from biting into the mouth of the urchin and eventually through the shell into its insides. It is difficult to imagine how a supposedly unintelligent fish was able to develop this behavior.

Another interesting predator is the sea otter which also uses its brain to overcome the formidable defenses of the urchin. Along the coast of California, otters have been observed floating on their backs breaking open the spiny test of the urchin, using rocks as tools for pounding.

Another less dramatic predator upon the

urchin is man. Man's relationship to the urchin is paradoxical. Some people spend considerable effort seeking the urchin as a gastronomic delicacy, while other people detest the urchin, and wantonly kill it with chemicals and hammers. Many Europeans (particularly the Italians and French) and the Japanese savor urchin gonads.

A sea otter (above) *swims up to purple sea urchin, next to giant kelp, to crack it open and eat it.*

Rubberlip sea perch (right) *feed on remains of sea urchin lying split open on ocean floor.*

Belugas v. Orcas

These handsome beluga, or white, whales of the Arctic Ocean are as successful as dolphins in their cold realm but they recognize and fear one old enemy—the orca, sometimes called the "killer whale." Scientists are using this fear to reduce predation by belugas on the salmon of the Canadian Pacific coast. In one experiment conducted in Bristol Bay, Alaska, orca sounds recorded on tape, and played back into the river through an appropriate underwater speaker, repulsed almost 500 belugas attempting to go up the river to prey upon young salmon. The white porpoises, during the three week period when the salmon are migrating down

> "Belugas recognize and fear one old enemy—the orca. Scientists are using the fear to reduce predations by belugas on the salmon off the Pacific coast of Canada."

the river, enter the waters twice daily to feed. During six different times of orca playbacks, the belugas were observed immediately to turn downstream and swim rapidly out to sea. On the seventh trial, the pod of whales separated and swam out, the eighth playback induced them to swim to the farside of the river and finally on the tenth session, all whales continued up the river, but on the farside of an adjacent sandbar.

Obviously the belugas eventually learned that the sounds were not from a live orca.

Belugas. *Scientists are now able to capitalize on the beluga whale's fear of the orca. By playing the sounds of the orca into a river, they have repulsed the beluga from feeding on salmon.*

However, the scientists feel that this deterrent will be effective during the short critical periods when the greatest number of salmon migrate in the rivers. It will be interesting to see how many of the whales will remember man's clever deception from year to year. It may be that in a near future the beluga population will ignore the orca calls entirely.

281

Sharks v. Dolphins

In the deadly but careful game of shark v. dolphin, the mammals are bound to win. But not always—as here we see a shark devouring the remains of a dolphin. The greater intelligence and vitality of the dolphin gives him the advantage. In great number, sharks always trail packs of dolphin waiting for a dropout—an ill animal or a young one—to fall behind the rest of its group. Then the shark moves in to devour the dolphin. Actu-ally the sharks are opportunists and never attack in the sense of a tuna stalking sardines. The dolphin is not a vicious warrior either. It surely does not search the seas for sharks simply for the joy of killing, but it can and does frequently get rid of sharks for safety reasons. Turning on the speed, they slam beak-first into the gill area of the soft lower abdomen of the sharks, their most vulnerable spots. The dolphins' beaks—their pointed jaws—are efficient weapons for such blows.

Several marine laboratories are studying the shark-dolphin relationship in the hope of making the dolphin's behavior useful to man. Experiments on lemon sharks and bottlenose dolphins have shown that if given the choice sharks will avoid dolphins. Then the researchers have been training dolphins to make sure that they could be used for shark control. One dolphin has been taught to ward off sharks in captivity on command from a sonic device. The dolphin, on cue, will chase and hit the shark. Soon the scientists will

The unfit. Despite the fact that dolphins usually win in battles against sharks, sharks often wait for an unfit dolphin to drop out of a group so they can devour him.

conduct these experiments in the open sea, hopefully employing dolphins to defend divers from sharks. Someday such trained dolphins may help oceanauts by acting as watchdogs around undersea habitats or they may police coastal beaches warding off sharks and protecting swimmers.

283

Chapter IX. A Time for Peace

For living dots and stars and germs
For jelly barrels, feather worms
For buds and bugs and shrimp and larvae
For clams and crabs and anemone

        With sham and poison, bites or gulps

Guns and armor or suction cups
Mean tricks and struggles—sly escapes

        Only to own or eat or mate.

In aimless drift, for pulsing clouds
On ocean floor for humble crowds
Which toil and moil in water winds
There is no break from fear and hunger.

In frantic feasts at dawn and dusk
Wings, fins and flippers bold and shy
Glance off and back their wakes and gleams

        When fangs and beaks have torn their spoils
        A fragile peace spreads in the Sea.
        Longing for more exotic foe
        The thoroughbred, masters of Space.

Harry quicksilver shapes that wing out of the brine
Or sound into the dark for iridescent prey.
Entrenched in oblivion denizens of the gloom
Squirt up to brighter realms and rape a beam of sun.

        The Ocean Lords with time to spare
        Display the Spirit of the Sea.

Index

ILLUSTRATIONS AND CHARTS:

Sy and Dorothea Barlowe—34–35, 118–119; Howard Koslow—21, 22, 23, 24, 25, 26-27, 33, 40, 41, 42, 52, 53, 62, 79, 96–97, 99, 116–117, 127, 173, 177.

PHOTO CREDITS:

Chuck Allen—166 (top); Ken Balcomb—36–37, 104–105, 241; John Boland—39, 179, 191 (bottom); Tony Chess—230; Jim and Cathy Church—226, 227; Ben Cropp—31, 91, 144-145, 158, 163, 229; David Doubilet—24 (top), 134–135, 159, 164, 168, 169, 184, 185, 207, 209, 233, 243, 261, 262–263, 264; Jack Drafahl, Brooks Institute of Photography—186, 239, 248, 256, 270; F. Ferro, M. Grimoldi, Rome—139; Freelance Photographers Guild: Alpha—12, Shayne Anderson—69, Ron Church—55, P. Colin—117 (top left), 259, James Dutcher—198–199, 203, 212, Robert Evans, Western Marine Labs—137, Robert B. Evans—191 (bottom), FPG—101, 214, Bob Gladden—66, 67, L. Grigg—231, Jerry Jones—235, 257, Stan Keiser—222-223, 265, Tom Myers—68, 110-111, 153, 178, 206, Chuck Nicklin—73, 89, 92–93, 133, 171, 194, 204, 213, 250, 266–267, R. Panouska—78–79, Siebe Rekker—95, John Stormont—53, 136, 260, A. B. Trail—63, 90, 113, 190 (bottom), 192, 211, 215 (bottom), Western Marine Laboratory—156, 180–181, 183, 224–225, 252, 279, John Zimmerman—85, 98; George Green—141; H. Hansen, Aquarium Berlin—251; Edmund Hobson—24, 50, 237; Dr. C. Scott Johnson, Naval Undersea Center, San Diego—172, 282–283; Holger Knudsen, Marine Biological Laboratory, Helsinger, Denmark—64, 218; Don Lusby, Jr.—280–281; Maltini–Solaini, M. Grimoldi, Rome—70, 71 (top), 100, 242, 253; Dr. Charles Mather—161; Bill McDonald—17; Jack McKenney, Skin Diver Magazine—47, 52, 57, 160; Richard Murphy—121, 284; Naval Undersea Center, San Diego—45, William E. Evans—106–107, John C. Sweeney—123; Chuck Nicklin—245; Carl Roessler—32, 46, 126–127, 188, 189, 193, 228, 272; George X. Sand—154; Dr. Paul R. Saunders, University of Southern California—176–177; John B. Schoup—234, 274–275; Sea World, Inc., San Diego, California—108–109; Tom Stack & Associates: Douglas Baglin—165, Ron Church—74–75, 117 (top right), 125, Bill DeCourt—246–247, Dr. E. R. Degginer—155, 221, LeRoy French—65, Warren Garst—268-269, M. J. Gilson—28, Ben Goldstein—82 (right), Richard F. Gunter—14, Dave LaTouche—271, Jack McKenney—138, Tom Myers—238, Tom Stack—19, William Stephens—86, 97 (top), 120, 208, 220; Submarine Flotilla One, Public Affairs Office, San Diego, California—128–129; Joe Thompson—157; Paul Tzimoulis—48, 72, 77, 83, 84; U.S. National Marine Fishery Service—165 (bottom); Don Wobber—219; Ed Zimbelman—147, 210.